Calendars and Years

Astronomy and Time in the Ancient Near East

Calendars and Years

Astronomy and Time in the Ancient Near East

edited by
John M. Steele

Oxbow Books

Published by
Oxbow Books, Oxford, UK

ISBN 978-1-84217-302-2

A CIP record for this book is available from the British Library

This book is available direct from

Oxbow Books, Oxford, UK
(Phone: 01865-241249; Fax: 01865-794449)

and

The David Brown Book Company
PO Box 511, Oakville, CT 06779, USA
(Phone: 860-945-9329; Fax: 860-945-9468)

or from our website

www.oxbowbooks.com

Printed and bound in Great Britain by
Hobbs the Printers Ltd.
Totton, Hampshire

Contents

Preface

The ordering of time by the use of a calendar is an essential tool in the writing of history. In order to construct a chronological story of the past it is necessary to understand how calendars used by different historical cultures operated and interlink with one another. The operational rules of many ancient calendars are now well understood by modern scholars and several handbooks for converting between dates give in ancient sources to the familiar Julian BC/AD calendar are available. However, very few historians understand how our present understanding of ancient calendars has been achieved. In particular, it is often far from clear what assumptions underlie the date-conversion tables presented in handbooks. By and large, current knowledge of an ancient calendar relies upon the accumulated work of scholars over several decades, or even centuries. In the process it is not always clear whether unjustified assumptions have crept into the accumulated understanding of an ancient calendar's operation and then been taken as fact by later scholars. It was in order to address this issue that Henry Zemel (CAENO Foundation) and I decided to organise a session on "Calendars and Years" at the Seventh Biennial History of Astronomy Workshop held at Notre Dame University on 8 July 2005.

Six of the papers in this book come directly from the Notre Dame meeting. They address the development and use of calendars in ancient Egypt and Mesopotamia, the interaction between astronomy and the Egyptian and Mesopotamian calendars, and the history of the modern understanding of these calendars. In addition, A. Jones generously contributed a paper on stellar and zodiacal date-reckoning in the Greco-Roman world. I had originally hoped to expand the coverage of the volume to include calendars and calendrical astronomy in other parts of the world (in particular China, India and Mesoamerica), but in order to publish the book without too great a delay after the meeting this was not possible. Perhaps a second "Calendars and Years" meeting could rectify these omissions.

This volume would not have been possible without the support and encouragement of Henry Zemel, who suggested the topic in the first place and proposed several of the speakers. I would like to thank him for his enthusiasm for this project. I would also like to thank the CAENO Foundation for financial support; Matthew Dowd and the organisers of the Notre Dame workshop for accepting the "Calendars and Years" session into their programme and for hosting an extremely successful conference series; Tom Boiy, Eleanor Robson, Pauline Russell and Sarah Symons for help with editing and/or producing figures; and the Royal Society for its support of my work through the award of a Royal Society University Research Fellowship.

John M. Steele
April 2007

A Star's Year:
The Annual Cycle in the Ancient Egyptian Sky

Sarah Symons

Introduction

This paper discusses the yearly cycle of a star in the Egyptian sky based on the evidence presented in various types of 'star clocks' and astronomical diagrams. The relationship between these astronomical representations and the civil calendar are explored, including the cycle of updates to the diagonal star clock tables proposed by Neugebauer and Parker. Content and classes of diagonal star tables, the triangle and epagomenal column, and some aspects of the origin and function of the tables are discussed. Finally, key conclusions are summarised.

Sources and Definitions

Diagonal star tables are usually painted on the inside surface of wooden coffin lids which date from the IXth to XIIth dynasties. One exception is a fragment of a diagonal star table on a ceiling in the Osireion at Abydos which dates from the XIXth dynasty. *Egyptian Astronomical Texts Volume 1*[1] contains details of twelve diagonal star tables on coffin lids (labelled 'Coffin 1' to 'Coffin 12') plus the table from the Osireion.[2] Eight further sources are now known. All twenty-one sources (T1 to T12 and K0 to K7 plus an empty grid) are listed in Table 1.[3] All are from Asyut unless otherwise noted. Schematic diagrams for each of the sources which contain decans are given in Tables 9 to 28 as an appendix to this paper.

The star tables have in the past been known as 'star calendars'[4] but are currently called 'diagonal star clocks'. The term 'clock' is, however, problematic and misleading. Strictly, the word can only properly be used for a mechanical timekeeping instrument that sounds the time but, most importantly, the application of the word 'clock' to the star tables encourages the perception that the intended function of the tables was as hourly timekeeping devices. This perception has been questioned by many researchers.[5] Avoiding the word 'clock' would allow the tables to be viewed with a more open mind concerning their construction and purpose, so this paper will use the term *diagonal star table.*

A diagonal star table consists of a grid containing the names of individual stars and, perhaps, asterisms (groups of stars forming small patterns). The particular stars and asterisms which appear in the context of star tables and astronomical ceilings are known as *decans.* The set of decans within any one source is called a *decan list.* Decan lists vary somewhat, but some decans occur in many or most lists (for example decans called *ḥry-ib wiꜣ*, *ẖꜣw*, and *knmt* are regular members). Although we know the names of the decans, and in some cases can translate the names (*ḥry-ib wiꜣ* means 'in the centre of the boat') the locations of the decanal stars and their relationships to modern star names and constellations are not known. This is due

Label	N&P[a]	Coffin[b]	Name of owner, date, and provenance	Orderly?	Date Row?	Rows x Columns + List Columns	Vertical Band Order	Horizontal Strip Order
T1	Coffin 1 (I)	S1C	*Msḥt* IX-Xth dynasty	Y	Y	12 x 36+4	NFOS	FNO
T2	Coffin 2 (I)	S3C	*Ịt-ib* IX-Xth dynasty	Y	Y	12 x 32	NFOS	FNOS
T3	Coffin 3 (I)	S6C	*Ḫw-n-Skr* usurped by *Ḫty* IX-Xth dynasty	Y	N	12 x 20	NFOS	FNO
T4	Coffin 4 (I)	S1Tü	*Ịdy* date unknown	Y	Y	12 x 19	NFOS	FNO
T5	Coffin 5 (I)	S2Chass	*Mꜣꜥt* IXth-Xth dynasty	Y	Y	12 x 16	NFOS	FNOS
T6	Coffin 6 (II)	T3C	*ꜥꜣšyt* XIth dynasty, from Thebes	Y	N	12 x 36+3	SOFN	FNO
T7	Coffin 7 (II)	G2T	*Ịḳr* First Intermediate Period or XIth dynasty, from Gebelein	Y	Y	12 x 36+3	SOFN	NOFS
T8	Coffin 8 (II)	A1C	*Ḥḳꜣt* date unknown, from Aswan	Y	Y	12 x 36+3	SOFN	NOFS
T9		S1Hil	*Nḫt* XI-XIIth dynasty[c]	Y	Y	12 x 27	NFOS	FNOS
T10		S16C	Name and date unknown, probably from Asyut[d]	Y	N	12 x ?	?	?
T11		S2Hil	Name and date unknown, probably from Asyut[e]	Y	Y	10 x ?	SOFN	?
T12	Coffin 9 (III)	S3P	*Ḫw-n-Skr* usurped by *Nḫt* usurped by *Ḥnn* IX-Xth dynasty	Y	Y	12 x 21	NFOS	FNOS
K0			The sloping passage star table from the Osireion at Abydos[f]	Y	Y	12 x ?	-	-
K1	Coffin 10 (IV)	S9C	*Ṯꜣwꜣw* XIIth dynasty	N	N	8 x 23	SOFN	OF
K2	Coffin 11 (V)	S5C	*Ṯꜣwꜣw* date unknown	N	N	12 x 24	SFON	FN
K3	Coffin 12 (V)	S11C	*Šms* XIIth dynasty	Y	N	12 x 17	SFON	OS
K4		S#T	Name and date unknown[g]	Y	N	12 x 19	NFOS	FNO
K5		X2Bas	Name and date unknown, probably from Asyut[h]	Y	N	8 x 12	NFOS	OFN
K6			British Museum EA47605. Name and date unknown[i]	?Y	Y	13 x ?	-	-
K7			British Museum (no EA number). Name and date unknown[j]	?	?	?6 x ?	-	-
		T3L	*Sbk-ḥtp* (British Museum EA29570), from Thebes			6 x 40	SOFN	OF

[a] Designations used in Neugebauer and Parker (1960) where bibliographies for these sources can be found. Neugebauer and Parker's Group number follows in brackets.
[b] For coffin designations see Lesko (1979).
[c] Inv. Nr. 5999 in the Pelizaeus-Museum in Hildesheim. See Eggebrecht (1990), 58–61 (including plates) and Eggbrecht (1993) pl. 33 (pp. 41–43).
[d] See Locher (1998).
[e] See Locher (1998).
[f] Franckfort (1933).
[g] See Locher (1983).
[h] First published in Lapp (1985) then by Locher (1992).
[i] First published in Symons (2002b).
[j] First published in Symons (2002b).

TABLE 1. Diagonal star table sources.

to many factors, but key problems are the uncertainty surrounding the observation methods used to develop and populate the diagonal star tables and the criteria used to select decans (brightness, position, relationship with other stars, and so on).

The main body of the diagonal star table grid usually has twelve *rows* and thirty-six *columns*. Each column represents one of the thirty-six *decades* (10-day periods) which make up the twelve months of the civil year.

Table 2 shows the layout of a hypothetical, idealised diagonal star table. The main body of the table caters for 360 days, and is often headed by a *date row* listing the thirty-six decades. The numbers 1 to 36 and the letters A to K which appear in the main body of the table each represent a decan name. The decan names represented by the letters A to K are usually called

I Akhet First decade	1	2	3	4	5	6	7	8	9	10	11	12
Middle decade	2	3	4	5	6	7	8	9	10	11	12	13
Last decade	3	4	5	6	7	8	9	10	11	12	13	14
II Akhet First decade	4	5	6	7	8	9	10	11	12	13	14	15
Middle decade	5	6	7	8	9	10	11	12	13	14	15	16
Last decade	6	7	8	9	10	11	12	13	14	15	16	17
III Akhet First decade	7	8	9	10	11	12	13	14	15	16	17	18
Middle decade	8	9	10	11	12	13	14	15	16	17	18	19
Last decade	9	10	11	12	13	14	15	16	17	18	19	20
IV Akhet First decade	10	11	12	13	14	15	16	17	18	19	20	21
Middle decade	11	12	13	14	15	16	17	18	19	20	21	22
Last decade	12	13	14	15	16	17	18	19	20	21	22	23
I Peret First decade	13	14	15	16	17	18	19	20	21	22	23	24
Middle decade	14	15	16	17	18	19	20	21	22	23	24	25
Last decade	15	16	17	18	19	20	21	22	23	24	25	26
II Peret First decade	16	17	18	19	20	21	22	23	24	25	26	27
Middle decade	17	18	19	20	21	22	23	24	25	26	27	28
Last decade	18	19	20	21	22	23	24	25	26	27	28	29
III Peret First decade	19	20	21	22	23	24	25	26	27	28	29	30
Middle decade	20	21	22	23	24	25	26	27	28	29	30	31
Last decade	21	22	23	24	25	26	27	28	29	30	31	32
IV Peret First decade	22	23	24	25	26	27	28	29	30	31	32	33
Middle decade	23	24	25	26	27	28	29	30	31	32	33	34
Last decade	24	25	26	27	28	29	30	31	32	33	34	35
I Shemu First decade	25	26	27	28	29	30	31	32	33	34	35	36
Middle decade	26	27	28	29	30	31	32	33	34	35	36	A
Last decade	27	28	29	30	31	32	33	34	35	36	A	B
II Shemu First decade	28	29	30	31	32	33	34	35	36	A	B	C
Middle decade	29	30	31	32	33	34	35	36	A	B	C	D
Last decade	30	31	32	33	34	35	36	A	B	C	D	E
III Shemu First decade	31	32	33	34	35	36	A	B	C	D	E	F
Middle decade	32	33	34	35	36	A	B	C	D	E	F	G
Last decade	33	34	35	36	A	B	C	D	E	F	G	H
IV Shemu First decade	34	35	36	A	B	C	D	E	F	G	H	I
Middle decade	35	36	A	B	C	D	E	F	G	H	I	J
Last decade	36	A	B	C	D	E	F	G	H	I	J	K
Epagomenal days	A 25 13 1	B 26 14 2	C 27 15 3	D 28 16 4	E 29 17 5	F 30 18 6	G 31 19 7	H 32 20 8	I 33 21 9	J 34 22 10	K 35 23 11	L 36 24 12

TABLE 2. Idealised layout of a diagonal star table.

the *triangle decans*, after the shape they make in the table. The decans 1 to 36 can be called *ordinary decans*. The distinctive diagonal pattern created by the decan names distinguishes this type of table from the two other types of 'star clock' – the Ramesside star clock and the so-called 'transit star clock' in the *Book of Nut* (also present in the Osireion at Abydos).

The four columns on the left of the table contain a list of all decans: the ordinary decans are in the first three *list columns* (which may have served as a reference list of decans used in the table), and the eleven triangle decans A to K plus one extra triangle decan L in the final or *epagomenal column*.[6]

The table is divided into quarters by a horizontal *offering text* and a vertical *band* containing figures of deities associated with the sky.

Grouping Systems

Neugebauer and Parker extensively analysed twelve star tables (T1 to T8, T12, and K1 to K3) and remarked that most tables are in some way corrupted.[7] Copyists' mistakes are rife in the main part of the table and many of the sources are incomplete. Neugebauer and Parker identified five groups of diagonal star tables, which they labelled Group I (consisting of sources T1 to T5), Group II (T6 to T8), Group III (T12), Group IV (K1), and Group V (K2 and K3).

Neugebauer and Parker stated that three layout elements contributed to their classification system (date row, vertical band content, and offering strip content) and that if these factors were considered, the coffins 'readily group themselves into five lots'.[8] This is far from true. The major grouping factor is the content of the tables: the lists of stars used. We can for now simplify the consideration of decan lists by noting that tables either start with decans in the *ṯmꜣt* area or the *knmt* area. Detailed discussion of decan lists is left until the next section.

Neugebauer and Parker's Group I contains five coffins from Asyut, IX-Xth dynasty, all having *ṯmꜣt*-style decan lists. They may or may not have date rows, but always have the order Nut, Foreleg, Orion, Sirius[9] in the vertical band and Foreleg, Nut, Orion, and (sometimes omitted) Sirius in the offering strip.

Group II consists of three coffins which are *not* from Asyut, but like Group I tables date from the IX-Xth dynasty, may or may not have date rows, and have *ṯmꜣt*-style decan lists. The major factor which appears to distinguish Group II tables is the order of figures in the vertical band: Sirius, Orion, Foreleg, Nut. The order of offerings varies.

Groups III and IV each contain only one coffin. The Group III coffin differs from the Group I standard by slight variations in the decans used. Similarly, the Group IV coffin differs in content from the Group V coffins, but in this case the order of figures in the vertical band is also different.

Finally, the two Group V coffins are distinguished by provenance, the lack of a date row, order in the vertical strip (the same as Group II), and decan lists which start in the *knmt* area. The characteristics of the five groups are summarised in Table 3.

Any successful classification system must be capable of incorporating all objects in its field. Neugebauer and Parker's grouping scheme can be easily tested against this criterion, because eight new tables have been published since their system was devised.

Table 4 demonstrates that not one of the new sources can be securely located in Neugebauer and Parker's scheme. Three sources are not complete enough nor well enough documented to allow them to be grouped, but the remaining five sources establish very clearly that the layout features of date row, offering text, and vertical band do *not* support the group

Group	Date and provenance	Date Row?	Vertical Band Order	Horizontal Strip Order	First decans
I	IX-Xth dynasty, Asyut	Y or N	NFOS	FNO[S]	*ṯmꜣt*
II	?XIth dynasty, **not** from Asyut	Y or N	SOFN	Any	*ṯmꜣt*
III	IX-Xth dynasty, Asyut	Y	NFOS	FNOS	*ṯmꜣt*
IV	XIIth dynasty, Asyut	N	SOFN	OF	*knmt*
V	?XIIth dynasty, Asyut	N	SFON	Any	*knmt*

TABLE 3. Criteria for Neugebauer and Parker's groups, reconstructed from Neugebauer and Parker (1960). Epigraphical considerations are not taken into account.

Label	Coffin	Date and provenance	Date Row?	Vertical Band Order	Horizontal Strip Order	First decans	N&P Group
T9	*S1Hil*	XI-XIIth dynasty, Asyut	Y	NFOS	FNOS	*ṯmꜣt*	I or III, except that date is wrong
T10	*S16C*	Date unknown, ?Asyut	N	?	?	*ṯmꜣt*	Not enough known
T11	*S2Hil*	Date unknown, ?Asyut	Y	SOFN	?	*ṯmꜣt*	II, except for provenance
K4	*S#T*	Date unknown, Asyut	N	NFOS	FNO	*knmt*	NFOS + *knmt* = no match
K5	*X2Bas*	Date unknown, ?Asyut	N	NFOS	OFN	*knmt*	NFOS + *knmt* = no match
K6		Date unknown, Asyut	Y	-	-	*knmt*	IV or V – except for date row
K7		Date unknown, Asyut	?	-	-	*?knmt*	IV or V – not enough known
E1	*T3L*	Thebes	N	SOFN	OF	-	Perhaps II

TABLE 4. Fitting new sources into Neugebauer and Parker's groups.

system and are *not* features which can be used to distinguish trends or relationships in diagonal star tables.

Importantly, no table in Group IV or V has a date row. Neugebauer and Parker implied and Leitz stated explicitly[10] that 'later' star tables (indicated by *knmt*-type decan lists) would never have had a date row. Leitz even speculated that *knmt*-type tables started at a different point in the year from *ṯmꜣt*-type tables. The publication of British Museum object EA47605[11] disproves this theory and others which are based on it.[12]

It should be noted that no classification system has successfully incorporated many newly identified sources without modification.[13] The impracticality of the schemes has meant that no system has supplanted Neugebauer and Parker's grouping method and become accepted by other researchers. This is an indication that the number of examples of star tables which have survived is not sufficient for a meaningful classification system to be devised which incorporates *all* points of layout, epigraphy, place of origin, and age. Such systems tend to the reduction of groups into individual sources. Kahl's stemma[14] at least had the virtue of showing some evolutionary or taxonomical arrangement of the star tables.

With a growing number of star table sources, some organisational structure is undoubtedly useful. Drawing experience from the previous attempts at classification leads to the conclusion that decan lists are the key features of these tables.

Decan Lists in Diagonal Star Tables

Each diagonal star table which has survived presents us with a decan list. However, this decan list may be fragmentary or disordered or both. Errors can occur at each re-writing. Errors which may have been present in a 'master copy' can be seen across several star tables, but each table compounds these common errors with idiosyncrasies of its own. Even so, we can attempt to reconstruct decan lists and, while doing so, find that the tables display similarities and differences which group them naturally into two classes.

In the discussion above, it was noted that decan lists in diagonal star tables began either with the *ṯmꜣt* decans or the *knmt* decans. This means that the decans occupying the cells in the top right-hand corner of an orderly star table are either *ṯmꜣt ḥrt* and *ṯmꜣt ẖrt* (for *ṯmꜣt*-class[15] or '**T**' tables, of which there are twelve) or *tpy-ꜥ knmt* and *knmt* (for the eight *knmt*-class or '**K**' tables). Most tables preserve the top-right-hand area, making their class obvious. Those which are damaged or disordered, can still be placed into one of the two classes with certainty. The only source which cannot be placed in this system is the empty table T3L.

Sixty-eight[16] individual decans appear within the twenty **T** or **K** tables. Several observations which differentiate classes of decan lists should be noted:

1. *Wšꜣt bkꜣt* splits into two separate decans *wšꜣti* and *bkꜣti* in T12 and K1 to K6.
2. Conversely, *ḳdty* becomes (or perhaps is replaced by) *spty* and "*ẖnwy*" becomes *ẖnwy* (the writing differs) in K0 and K1, then the two new decans merge to a single decan *spty ẖnwy* in K2, K3, K4, and K5.
3. "Crew" changes writing to *sšmw* from K0.
4. *Tpy-ꜥ smd* is absent in all the **T** tables.
5. *Smd srt* contracts to *smd* (both writings appear in T12).
6. The triangle decans, in particular *sꜣbw*, will be discussed presently.
7. The decans in the *sꜣḥ* area are confused and problematic. T1 to T9 and T12 have three decans: *rmn ḥry*, *ꜥbwt*, *ẖrt wꜥrt*. *Rmn ẖry* (which we might expect as a pair with *rmn ḥry*) only appears at the top of the penultimate column of T1 between *rmn ḥry* and *ꜥbwt*.

Order of *ꜣẖwy* and *bꜣwy* Decans in T

An important point about decan order is not visible in Tables 5 and 6. Neugebauer and Parker made a significant change in order of decans in one area of the list, despite the order being consistently demonstrated in the **T** sources. The area in question contains potentially four decans: *tpy-ꜥ ꜣẖwy*, *imy-ḫt ꜣẖwy*, *ꜣẖwy*, and (omitted, but perhaps intended) *bꜣwy*. Neugebauer and Parker put them in the order *tpy-ꜥ ꜣẖwy*, *ꜣẖwy* (always omitted, but assumed intended), *imy-ḫt ꜣẖwy*, and *bꜣwy* (mistakenly written as *ꜣẖwy* throughout). Neugebauer and Parker stated[17] that '*imy-ḫt*' means 'following' and in order to support this translation stipulate that the decan written *ꜣẖwy* must be emended to *bꜣwy* throughout the main body of the tables, as the difference in writing between *bꜣwy* and *ꜣẖwy* is only a matter of the type of bird drawn (for *bꜣ* and for *ꜣẖ*).

There is some evidence that there should be four *ꜣẖwy*/*bꜣwy* decans as all four appear in K1. Unfortunately, in K1 the decans are in disarray, so cannot be used to decide order. *Bꜣwy* does not appear in any of the **T** class tables. In later decan lists, *ꜣẖwy* and *bꜣwy* are consistent and distinct members, so the restoration of the fourth decan is reasonably secure. Returning to the question of order, the surviving sources must be our primary guide. Here, there is not a good enough reason for Neugebauer and Parker's convoluted change. The translation issue is not evidence: *imy-ḫt* can mean 'in the entourage of' or 'accompanying'.[18] Neither rendering prevents *imy-ḫt 3ḫwy* from preceding *3ḫwy*. Additionally, we have no idea how the 'figures' which these star names represented were depicted or arranged.

The list columns also do not support Neugebauer and Parker's theory. In T1 two instances of *3ḫwy* appear, one before and one after *imy-ḫt 3ḫwy*. The second of these instances was probably a poorly-written *b3wy*. In T7 and T8, *3ḫwy* follows *imy-ḫt 3ḫwy* exactly as it does in the body of the table.

This detailed examination of the *3ḫwy* and *b3wy* area leads inevitably to the conclusion that there is nothing to be gained by changing the order of these decans. We also have seen that *b3wy* is a likely candidate for a 'missing' decan, which we will discuss below.

Order of *ṯm3t* Decans in T

T1, T6, T7, and T8 are our most complete sources, but each one contains just thirty-four ordinary decans in the main body of the table, where we would expect thirty-six. The thirty-fourth decan is followed by a repetition of the two *ṯm3t* decans, and then by the first triangle decan *smd rsy*. Repeating the *ṯm3t* decans is an odd mistake, given that this is not a single writing error but two diagonal lines of decans spanning the width of the table. That the list should be missing two ordinary decans is not exceptional, but that the mistake should be noticed and corrected mid-way through the table, so that the triangle decans, in the most 'difficult' part of the table, are correctly placed is thought provoking.

As the mistake of omitting two ordinary decans and of repeating the *ṯm3t* decans appears uniformly across the **T** tables, the error was almost certainly present in the master source or sources from which these tables were copied. This points strongly to all these tables, including the slightly differing T12, having a common ancestor. This ancestor may or may not have been a faulty copy of a 'perfect' original star table (possibly there never was such a table) but in reconstructing the relationships between the star tables and the circumstances and process of their development, identifying where in the table the mistake occurred is important.

The copyist, who perhaps did not fully understand the text on which he was working, may have started at the right-hand end of the table (a typical direction for writing), with *ṯm3t ḥrt* and *ṯm3t ḫrt* correctly placed, then continued working leftwards, diagonal by diagonal. He missed two decans without noticing, then, mid-table, at a point which indicates some understanding of the structure of the triangle decans (even though these are never marked in any way to distinguish them in the existing sources from ordinary decans), he corrected his error by inserting two full diagonals of *ṯm3t* decans before filling in the triangle. This process would mean that the first appearance of the *ṯm3t* decans is the correct placing, and the second is erroneous.

If instead the first pair of *ṯm3t* decans is incorrect, this could have happened in at least two ways. If the copyist worked from left-to-right, which is reasonable but not the more usual direction, then the two missing ordinary decans would only be noticed when all but the three top-right-hand corner cells of the table were yet to be filled. By placing two instances of *ṯm3t ḫrt* and just one instance of *ṯm3t ḥrt*, the mistake could be covered up. However, it might be that this small area was instead damaged or unreadable in the master copy. The same process of filling in three *ṯm3t* labels might apply, and the table could be completed working right-to-left instead.

NP	T		T1	T2	T3	T4	T5	T6	T7	T8	T9	T10	T11	T12
1	**35**	*ṯmꜣt ḥrt*	X	X	X	X	X	X	X	X	X			X
2	**36**	*ṯmꜣt ẖrt*	X	X	X	X	X	X	X	X	X	X		X
3	**1**	*wšt bkꜣt*	X	X	X	X	X	X	X	X	X	X		-
3a	**1a**	*wšꜣti*	-	-	-	-	-	-	-	-	-	-		X
3b	**1b**	*bkꜣti*	-	-	-	-	-	-	-	-	-	-		X
4	**2**	*ipḏs*	X	X	X	X	X	X	X	X	X	X		X
5	**3**	*sbšsn*	X	X	X	X	X	X	X	X	X	X		X
6	**4**	*ḫntt ḥrt*	X	X	X	X	X	X	X	X	X	X	X	X
7	**5**	*ḫntt ẖrt*	X	X	X	X	X	X	X	X	X	X	X	X
8	**6**	*ṯms n ḫntt*	X	X	X	X	X	X	X	X	X	X	X	X
9	**7**	*ḳdty*	X	X	X	X	X	X	X	X	X	X		X
10	**8**	*"ḫnwy"*	X	X	X	X	X	X	X	X	X	X	X	X
11	**9**	*ḥry-ib wiꜣ*	X	X	X	X	X	X	X	X	X		X	X
12	**10**	*"crew"*	X	X	X	X	X	X	X	X	X		X	X
13	**11**	*knm*	X	X	X	X	X	X	X	X	X		X	X
14	**12**	*smd srt*	X	X	X	X	X	X	X	X	X		X	X
14a	**12a**	*smd*	-	-	-	-	-	-	-	-	-		-	X
15	**13**	*srt*	X	X	X	X	X	X	X	X	X		X	X
16	**14**	*sꜣwy srt*	X	X	X	X	X	X	X	X	X		X	X
17	**15**	*ẖry ḫpd srt*	X	X	X	X	X	X	X	X	X			X
18	**16**	*tpy-ʿ ꜣḫwy*	X	X	X	X	X	X	X	X	X	X	X	X
20	**17**	*imy-ḫt ꜣḫwy*	X	X	X	X	X	X	X	X	X	X	X	X
19	**18**	*ꜣḫwy*	X	X	X	X	X	X	X	X	X	X		X
21	**19**	*bꜣwy*												
22	**20**	*ḳd*	X	X	X	X	X	X	X	X	X			X
23	**21**	*ḫꜣw*	X	X	X	X	X	X	X	X	X		X	X
24	**22**	*ʿrt*	X	X	X	X	X	X	X	X	X		X	X
25	**23**	*ẖry ʿrt*	X	X	X	X	X	X	X	X	X	X	X	X
26	**24**	*rmn ḥry*	X	X	X	X	X	X	X	X	X			X
27	**25**	*rmn ẖry*	L											
28	**26**	*ʿbwt*	X	X	X	X	X	X	X	X	X			X
29	**27**	*ẖrt wʿrt*	X	X	X	X	X	X	X	X	X			X
30	**28**	*tpy-ʿ spd*	X	X	X	X		X	X	X	X			X
31	**29**	*spd*	X	X	X	X		X	X	X	X			X
32	**30**	*knmt*	X	X	X	X		X	X	X	X			X
33	**31**	*sꜣwy knmt*	X	X	X			X	X	X	X			X
34	**32**	*ẖry ḫpd n knmt*	X	X				X	X	X	X			
35	**33**	*ḥꜣt ḫꜣw*	X	X				X	X	X	X			
36	**34**	*pḥwy ḫꜣw*	X	X				X	X	X	X			
1	**35**	*ṯmꜣt ḥrt*	X	X				X	X	X	X			
2	**36**	*ṯmꜣt ẖrt*	X	X				X	X	X	X			
A	**A**	*smd rsy*	X	X				X	X	X	X			
B	**B**	*smd mḥty*	X	X				X	X	X	X			
C	**C**	*nṯr ḏꜣ pt*	X	X				X	X	X				
D	**D**	*rmn ẖry*	X	X				X	X	X				
E	**E**	*ḫꜣw*	X	X				X	X	X				
F	**F**	*tpy-ʿ spd*	X	X				X	X	X				
G	**G**	*imy-ḫt spd*	X	X				X	X	X				
H	**H**	*ꜣḫwy*	X					X	X	X				
J	**I**	*ḫꜣw*	X					X	X	X				
K	**J**	*nṯr ḏꜣ pt*	X					X	X	X				
M	**K**	*pḥwy sꜣbw*	X					X	X	X				
L	**L**	*sꜣbw*						L		L				

TABLE 5. Decans in class **T** diagonal star tables. 'X' indicates a decan is present in the star table. 'L' indicates that the decan only occurs in the epagomenal or list columns of the table and '–' that the decan is not expected to occur. Grey shading represents parts of the decan list that are missing owing to loss or damage, rather than to omission.

Each explanation of the *ṯmꜣt* error implies that the person who was creating, copying, or repairing the 'ancestral source' of the *ṯmꜣt* group tables knew the difference between triangle decans and ordinary decans. The mistake was therefore made by an educated, knowledgeable person rather than by a tradesman working quickly to decorate and sell quantities of funerary equipment. It seems unlikely that the error occurred during the invention of the tables as it resembles a scribal rather than an experimental error.

Although there is no certain resolution of the dilemma of which pair of *ṯmꜣt* decans is correctly placed, the simplest interpretation, which requires the least adjustment or the fewest implied writing errors, is that the pair at the end of the ordinary decans is in the intended location.[19]

The next question about this decan list is the position and identity of the two missing decans. In the analysis above, one suggestion has already been raised: the two missing decans were at the head of the table and were lost when three cells were destroyed. This is unlikely. Comparing this decan list with others, in other star tables and in later astronomical ceilings, no decan list indicates that the order *ṯmꜣt ḥrt*, *ṯmꜣt ẖrt*, *wšꜣti* (or *wšt bkꜣt*) could be interrupted by two unknown decans in the sequence *ṯmꜣt ḥrt*, *ṯmꜣt ẖrt*, unknown decan A, unknown decan B, *wšꜣti*.

The most likely explanation is that two of the decans which appear intrusively in the list columns were the missing ordinary decans. One, discussed above, is *bꜣwy*. We have also already mentioned the other candidate, *rmn ẖry*, in point 7 above. Both these decans are very reasonable suggestions for the missing ordinary decans, with perhaps the caveat that *rmn ẖry* adds to an already over-burdened *sꜣḥ* area of the list.

Order of Decans in T

The list **T** of thirty-six ordinary decans can now be reconstructed with *bꜣwy* (19) and *rmn ẖry* (25) proposed as the two missing ordinary decans:

1) *wšt bkꜣt*
2) *ipḏs*
3) *sbšsn*
4) *ẖntt ḥrt*
5) *ẖntt ẖrt*
6) *ṯms n ẖntt*
7) *ḳdty*
8) "*ẖnwy*"
9) *ḥry-ib wiꜣ*
10) "*crew*"
11) *k(n)m*
12) *smd srt*
13) *srt*
14) *sꜣwy srt*
15) *ẖry ẖpd srt*
16) *tpy-ʿ ꜣẖwy*
17) *imy-ẖt ꜣẖwy*
18) *ꜣẖwy*
19) *bꜣwy*
20) *ḳd*
21) *ẖꜣw*
22) *ʿrt*
23) *ẖry ʿrt*
24) *rmn ḥry*
25) *rmn ẖry*
26) *ꜣbwt*
27) *ẖrt wʿrt*
28) *tpy-ʿ spd*
29) *spd*
30) *knmt*
31) *sꜣwy knmt*
32) *ẖry ẖpd n knmt*
33) *ḥꜣt ẖꜣw*
34) *pḥwy ẖꜣw*
35) *ṯmꜣt ḥrt*
36) *ṯmꜣt ẖrt*

The decan list used in T12, a disordered source, differs primarily in that *wšt bkꜣt* is split into two distinct decans *wšꜣti* and *bkꜣti* (1a and 1b in Table 5) and that *smd srt* became simply *smd* (12a). The reason for the *wšt bkꜣt* split is not clear from this sole source, but the same feature appears in the second class of diagonal star tables, discussed below. T12 could indicate that another class of tables with a decan list similar but not identical to **T** existed. If more sources appear, the formation of a sub-class **Ta** might be supportable, but at present there is insufficient data: the differences in T12 could be unique to that table.

Order of Decans in K

No table complete up to the list columns is present in our second group of sources: those which appear to have *knmt* decans at the head of their lists. Several sources have unusual numbers of rows, some are jumbled, others are fragmentary. A single ancestor like that of the **T** tables is not as apparent. Generally, the eight **K** sources display the same major features in their decan lists and seem to have been designed from a similar source. The Orion area, where the evidence from extant sources begins to fall away, holds the greatest uncertainty. A possible reconstruction of **K** is:

1) *tpy-ʿ knmt*
2) *knmt*
3) *ẖry ḫpd n knmt*
4) *ḥꜣt ḏꜣt*
5) *pḥwy ḏꜣt*
6) *ṯmꜣt ḥrt*
7) *ṯmꜣt ẖrt*
8) *wšꜣtỉ*
9) *bkꜣtỉ*
10) *sšpt*
11) *tpy-ʿ ḫntt*
12) *ḫntt ḥrt*
13) *ḫntt ẖrt*
14) *ṯms n ḫntt*
15) *spty ḫnwy*
16) *ḥry-ỉb wỉꜣ*
17) *sšmw qr šsmw*
18) *knm*
19) *tpy-ʿ smd*
20) *smd*
21) *srt*
22) *sꜣwy srt*
23) *ẖry ḫpd srt*
24) *tpy-ʿ ꜣḫwy*
25) *ꜣḫwy*
26) *bꜣwy*
27) *ḫntw ḥrw*
28) *ḫntw ẖrw*
29) *ḳd*
30) *sꜣwy ḳd*
31) *ḫꜣw*
32) *ʿrt*
33) *rmn ḥry sꜣḥ*
34) *rmn ẖry sꜣḥ*
35) *rmn sꜣḥ*
36) *spd*

K0 and K1 retain certain distinct features, and it is possible that (again, with the support of new sources) they could form a sub-class **Ka**. **Ka** would include *spty* (15a in Table 6) and *ḫnwy* (15b) as separate decans. More speculatively, *sšpt* may be missing from the ordinary decans whereas *ỉmy-ḫt ꜣḫwy* might be present as an ordinary decan.

Astronomical Diagrams and Diagonal Star Tables

Neugebauer and Parker chose to include astronomical diagrams in their discussion of decan lists, in particular the list in the ceiling of Senenmut. There are several links between star tables and astronomical diagrams over and above the obvious celestial theme. Astronomical diagrams survive from later periods in history but there is some evidence that they evolved from representations on coffin lids.[20] Most astronomical diagrams are found in funerary contexts, in functionally identical locations to star tables: undersides of lids and ceilings. Most astronomical diagrams also contain decan lists which start with *tpy-ʿ knmt*, the same decan which leads the second group of star tables.

The decan lists in typical astronomical diagrams[21] display less variation than those in the two groups of star tables. Table 7 shows decan lists taken from astronomical diagrams in the New Kingdom. The astronomical diagrams add a further element to the list which does not appear clearly in any of the surviving second group sources: the triangle. The triangle is never complete, but we can gather that each astronomical diagram contained a fairly consistent set of triangle decans with a maximum of seven members.[22] We must also note that one family of decan lists consistently contains thirty-nine decans. Three of these could possibly be triangle decans. No decan list in a surviving New Kingdom astronomical diagram contains exactly thirty-six ordinary decans.

NP	K		K0	K1	K2	K3	K4	K5	K6	K7
31a	**1**	*tpy-ꜥ knmt*		X	X	X			X	[X]
32	**2**	*knmt*		X	X	X		X	X	X
34	**3**	*ḫry ḫpd n knmt*		X	X	X	X	X	X	X
35a	**4**	*ḥꜣt ḏꜣt*		X	X	X	X	X	X	
36a	**5**	*pḥwy ḏꜣt*		X	X	X	X	X	X	
1	**6**	*ṯmꜣt ḥrt*		X	X	X	X	X	X	
2	**7**	*ṯmꜣt ḫrt*		X	X	X	X	X	X	
3a	**8**	*wšꜣti*		X	X	X	X	X	X	
3b	**9**	*bkꜣti*		X	X	X	X	X	X	
4a	**10/C?**	*sšpt*			X		X	X	X	
5a	**11**	*tpy-ꜥ ḫntt*		X	X	X	X	X	X	
6	**12**	*ḫntt ḥrt*		X	X	X	X	X	X	
7	**13**	*ḫntt ḫrt*		X	X	X	X	X	X	
8	**14**	*ṯms n ḫntt*		X	X	X	X	X		
9a	**15a**	*spty*	X	X	-	-	-	-		
9b	**15**	*spty ḫnwy*	-	-	X	X	X	X		
10a	**15b**	*ḫnwy*	X	X	-	-	-	-		
11	**16**	*ḥry-ib wiꜣ*	X	X	X	X	X	X		
12a	**17**	*sšmw*	X	X	X	X	X	X		
13	**18**	*knm*	X	X	X	X	X	X		
13a	**19**	*tpy-ꜥ smd*	X		X	X	X	X		
14a	**20**	*smd*	X	X	X	X	X	X		
15	**21**	*srt*	X	X	X	X	X	X		
16	**22**	*sꜣwy srt*		X	X	X	X			
17	**23**	*ḫry ḫpd srt*		X	X	X	X			
18	**24**	*tpy-ꜥ ꜣḫwy*		X	[X]	X	X			
19	**25**	*ꜣḫwy*		X	X		X			
21	**26**	*bꜣwy*		X	X	X	X			
21a	**27**	*ḫntw ḥrw*		X	X	X	X			
21b	**28**	*ḫntw ḫrw*		[X]	X	X	?			
22	**29**	*ḳd*		X	X	X	X			
22a	**30**	*sꜣwy ḳd*		X		X	X			
23	**31**	*ḫꜣw*			X		X			
24	**32**	*ꜥrt*			X					
26a	**33**	*rmn ḥry sꜣḥ*			X					
27a	**34**	*rmn ḫry sꜣḥ*			X					
27b	**35**	*rmn sꜣḥ*			X					
31	**36**	*spd*			X					
31b	**A?**	*štwy*			X					
4	**B?**	*ipḏs*			X					
4a	**10/C?**	*sšpt*			X		X	X	X	
5	**D?**	*sbšsn*			X					
20	**E?**	*imy-ḫt ꜣḫwy*		X						

TABLE 6. Decans in class **K** diagonal star tables. 'X' indicates a decan is present in the star table. '[X]' or '?' indicates a damaged or possible reading and '–' that the decan is not expected to occur. Grey shading represents parts of the decan list that are missing owing to loss or damage, rather than omission.

The extent to which we can rely on the later astronomical diagrams to help us construct a decan list for the second class of diagonal star tables is debatable. Since *Egyptian Astronomical Texts*, what Neugebauer and Parker saw as a clear link between the Senenmut list and their Group IV and V star tables has been clouded and the issue must be revisited.

In astronomical diagrams, *sšpt* is a well-attested triangle decan. It is never one of the ordinary decans. Among Neugebauer and Parker's thirteen diagonal star clock sources, *sšpt* appeared solely in the disorderly K2. Considering the astronomical diagrams as guides, Neugebauer and Parker plausibly labelled *sšpt* as an out-of-place triangle decan. However, *sšpt* has since appeared as an ordinary decan in three of the new sources (K4 and K5, which

Senenmut's tomb KV[1] tombs: Ramesses VI, Ramesses VII (N)[2], Ramesses IX (N)	Karnak water-clock Mortuary temples: Ramesseum, Medinet Habu	Abydos temples: Seti I, Ramesses II KV tombs: Seti I, Tausert, Ramesses VI, Ramesses VII (S), Ramesses IX (S)
tpy-ꜥ knmt	tpy-ꜥ knmt	tpy-ꜥ knmt
knmt	knmt	knmt
ẖry ẖpd knmt	ẖry ẖpd knmt	ẖry ẖpd knmt
ḥꜣt ḏꜣt	ḥꜣt ḏꜣt	ḥꜣt ḏꜣt
pḥwy ḏꜣt	pḥwy ḏꜣt	pḥwy ḏꜣt
ṯmꜣt ḥrt	ṯmꜣt ḥrt	tmꜣt ḥrt
ṯmꜣt ẖrt	ṯmꜣt ẖrt	tmꜣt ẖrt
wšꜣti	wšꜣti	wšꜣty bkꜣty
bkꜣti	bkꜣti	sbꜣw mḥw
tpy-ꜥ ẖntt	tpy-ꜥ ẖntt	tpy-ꜥ ẖntt
ẖntt ḥrt	ẖntt ḥrt	ẖntt ḥrt
ẖntt ẖrt	ẖntt ẖrt	ẖntt ẖrt
ṯms n ẖntt	ṯms n ẖntt	ṯms n ẖntt
sꜣpti ẖnwy	sꜣpti ẖnwy	sꜣpti ẖnwy
ḥry-ib wiꜣ	ḥry-ib wiꜣ	ḥry-ib wiꜣ
sšmw	sšmw	šsmw
knmw	knmw	knmw
tpy-ꜥ smd	tpy-ꜥ smd	tpy-ꜥ smd
smd	smd	smd
sit	sit	srt
sꜣwy sit	sꜣwy sit	sꜣwy srt
ẖry ẖpd srt	ẖry ẖpd srt	ẖry ẖpd srt
tpy-ꜥ ꜣẖ(wy)	tpy-ꜥ ꜣẖ(wy)	tpy-ꜥ ꜣẖwy
ꜣẖwy[3]	ꜣẖwy	ꜣẖwy
bꜣwy	bꜣwy	bꜣwy
ẖntw ḥr(w)/ḥrt	ẖntw ḥr(w)	ẖnt(w) ḥrw ḥry-ib ẖnt(w)
(ẖntw) ẖrw/ẖrt	(ẖntw) ẖrw	ẖnt(w) ẖrw
ḳd	ḳd	ḳd
sꜣwy ḳd	sꜣwy ḳd	sꜣwy ḳd
ẖꜣw	ẖꜣw	ẖꜣw
ꜥrt	ꜥrt	ꜥrt
ḥry rmn sꜣḥ	ḥry rmn sꜣḥ	iwn sꜣḥ[4] rmn ḥry sꜣḥ
ẖry rmn sꜣḥ	ẖry rmn sꜣḥ	msḏr sꜣḥ[4] rmn ẖry sꜣḥ
sꜣḥ	rmn sꜣḥ	ꜥ sꜣḥ[4] sꜣḥ
spdt	spdt	spdt
štwy	štwy	štwy siꜣtw
nsrw	nsrw	nsrw
šspt	šspt	šspt
ipds	ipds	nhs
sbšsn	sbšsn	sbšsn
nṯr wꜣš	nṯr wꜣš	nṯr wꜣš

[1] Valley of the Kings
[2] (N) and (S) indicate the north and south parts of paired astronomical diagrams
[3] Missing in the KV tombs.
[4] Possibly triangle decans.

TABLE 7. Decan lists from New Kingdom astronomical diagrams. Triangle decans (lower section of the table) are always separated grapically from 'ordinary' decans in astronomical diagrams. Full details of all of these diagrams can be found in Neugebauer and Parker (1969).

are orderly tables, and K6 which is only a fragment). The weight of evidence now places *sšpt* as an ordinary decan.

Neugebauer and Parker focussed on the similarities between astronomical diagram decan lists and diagonal star clock decan lists. With *sšpt* now forcing a divide between the two areas, it is worth noting that decan lists set in an astronomical diagram display other characteristics which differentiate them from star table lists. These include their format (a list instead of a table), the association of names of 'deities of the decans' who never appear in diagonal star tables, the addition of humanoid figures representing these deities, other figures representing constellations, and numbers of 'stars', circles or star-symbols, relating to each decan name.

While it is still possible to say that astronomical diagram decan lists resemble **K** more than **T**, and that astronomical diagrams still are the best evidence for the composition of the **K** triangle, the association must now be considered a secondary one.

The T and K Class System

The sources have been divided into two traditions or classes, **T** and **K**, each of which retains a degree of flexibility, rather than stringently discriminating between sources which display slight dissimilarities. The lists **T** and **K** are not definitive, but are based on evidence derived from the diagonal star tables, but also drawing a little on astronomical diagrams. During its development,[23] the system has successfully incorporated the sources published recently.

Schematic diagrams of all twenty diagonal star tables which contain decans are presented here in Tables 9 to 28 at the end of this paper, with the decans numbered in the **T** and **K** system as described in Tables 5 and 6. Using the **T** and **K** system compares favourably with Neugebauer and Parker's decan numbering system, especially in the **K** class tables, where clarity and recognition of patterns is improved by the new system.

The **T** and **K** class system reflects family trends present in the creation of the sources under scrutiny rather than following a traditional 'ancestral tree' approach and, it is hoped, will be successful in incorporating further sources.

Significance of the Triangle

The triangle presents a major challenge in diagonal star table research. The ordinary decans represent an 'ideal state' and hint at the perfection of a group of thirty-six stars which should be sufficient for star table purposes. The inclusion of the triangle decans marks the impact of the real world—the physical model we now understand as the relationship between our planet, our sun, and the star sky—on the theoretical concept of the star table. If the tables were merely representations of a pattern or an ideal, the triangle would not be necessary and would certainly not have been invented. However, the triangle does exist, and there are only two ways of explaining why it is there: either because its existence was derived from an understanding of star motion and year length (that is, a theoretical model which allowed someone to calculate that 'star 1' would definitely not suffice after 'star 36'; an unusual idea in the context of ancient Egyptian thought in general but not impossible), or, more simply, because the table was based on observation of the real stars. For these reasons, the present author views the creation of the triangle as one of the most important and suggestive astronomical activities in ancient Egypt.

The triangle is a direct result of just two factors: that the tables were based on ten-day periods, and that it takes a star 23 hours, 56 minutes, and about 4 seconds to perform one complete (apparent) revolution. This period is called 'a sidereal day' and is a fundamental astronomical concept. We can model decanal stars by imagining ideal observing conditions and stars being available exactly where we want them, but making no further assumptions about the way the tables were developed and used, to demonstrate the relationship between the 365–day civil year and the triangle.

We imagine making an observation of a certain star labelled 'star 1' (we can imagine a star on the celestial equator, but another location makes no difference to the outcome) at a certain place in the sky (say a point relative to a mountain peak when viewed from a particular location) at a known time of day on, say, the first day of the year (I Akhet 1). From this notional fixed point in time and space, we continue to make similar observations at the ten-intervals specified by the diagonal star tables. The next observation, on I Akhet 11, will show that a new star ('star 2') will be in our chosen location at the same time of day. To quantify the distance between the two stars '1' and '2', we deduce that a star moves at a certain speed, taking 23 hours, 56 minutes, and about 4 seconds to move the full 360 degrees of a circle to return to our chosen observation point. Over the course of ten days, we calculate that the angular distance between the two stars will be about 9.9 degrees.[24] As we progress through the year, further stars will be found at ten-day intervals, but the angular distance between two consecutive stars will always be 9.9 degrees. When the 36th star is seen it will be IV Shemu 21, 350 days after our observations started, and 345.0 degrees total angular difference from 'star 1'. Ten days later, which in the Egyptian civil year would be the first epagomenal day, the total angular difference would be 354.8 degrees (that is, well after 'star 36' but somewhat before 'star 1'). The imagined star at this position, 'star A' is analogous to the first triangle decan which we see in the existing diagonal star tables. Five days after that, on I Akhet 1, the total angular difference would be 359.8 degrees, just a quarter of a degree away from 'star 1'. 'Star 1' is a good enough fit to use and so the table would work for the next year adequately. This model matches the top row of an Egyptian diagonal star table. The second row would follow a similar pattern, but would of course add another triangle decan 'star B' on the first epagomenal day. The ideal location of 'star B' would be at a total angular difference of 364.7 degrees (= 4.7 degrees past 'star 1', that is, between 'star 1' and 'star 2'). The other 'triangle' decans follow similarly.

The quarter of a degree difference between the 'ideal' star for the new I Akhet 1 and the old 'star 1' reflects the fact that 365 days is only an approximation to the length of the solar year. It would take several years before tables based on a functionally similar model became observationally inaccurate.

Despite the existence of the triangle and the triangle decans in the earliest diagonal star tables, it is however clear that the 'ideal' of thirty-six decans, the ordinary decans, had already become entrenched. Although the triangle was never divided graphically in the main body of the table, a label in the final column of T1, at the end of the list of decans, reads "*the total of those who are in their places … the gods of the sky* [*decans*]*: 36*".[25] The concept of the distinct 'thirty-six decans' lasted until the Greco-Roman Period.

The Epagomenal Column

On page one of *Egyptian Astronomical Texts* Volume 1 (which throughout treated the tables as corrupt copies of a functioning ancestor) it is conjectured that originally the diagonal star clock had thirty-seven useful columns. The thirty-seventh column represented (or was used in) the five epagomenal days at the end of the Egyptian civil year. At some point, the other list columns were added, giving this area of the table conjecturally the dual function of listing all the decans used and also being serviceable during the epagomenal days.

Evidence for the epagomenal days' explicit presence in the tables is contained in two of Neugebauer and Parker's thirteen sources, Coffins 7 and 8, here designated T7 and T8, in the form of a short text or label placed in the date row above the list columns.

T8 has the label showing clearly in the cells above the first two list columns: *rnpt* 5 *ḥrw* 'five (days) upon the year' followed at a short distance by the festival determinative sign 𓎰. The number 'five' is written cursively. The label in T7 in the plates of Neugebauer and Parker (1960) is not so clear. Neugebauer and Parker, following de Buck,[26] stated that this lid had been lost. This is not correct: the lid is on display (as viewed in November 2004) in the Museum of Ancient Egypt in Turin. Traces of the label, written in red on a richly-coloured wood, are now barely visible. However, photographic records[27] (perhaps taken using coloured filters to bring out the contrast) show that the label, in this case broken up by the large star motifs interspersed between the columns of text across the table, was of a similar format to that of T8.

However, the epagomenal label is not necessary to prove that these tables were based on a 365-day year. Using our model described above, we can test what would happen if the civil year had, for example, only 360 days. Everything would be the same up to and including the observation on IV Shemu 21. Ten days later, the '360-day-year' would begin again, but 'star 1' would be impossible to use, being about five degrees away from where it was needed. We would still therefore need to designate a new star 'A', but this time it would need to replace 'star 1' at the head of the table. Ten days later, the total angular difference would be 364.7 degrees, and again, a new star 'B' (as before, half way between 'star 1' and 'star 2') would be needed. This would continue across the whole table. In other words, the usable lifetime of a table created under these conditions would be only one year. Any ability to 'predict' star motions or 'record' a continuous state would be lost. The tables would probably not have existed in their present form at all. The 365-day Egyptian civil year is implicit in the tables and is responsible for their format and content.

These findings are true whether or not the epagomenal column was ever a functioning part of the table. The labels lend support to the theory that the area was functionally associated with the epagomenae, but do not settle the question unequivocally. For example, both the remaining labels span more than one list column and are therefore not solely associated with the triangle decans which the hypothetical thirty-seventh column would contain. Neugebauer and Parker theorised[28] that the other three list columns (those containing the list of thirty-six ordinary decans) represent an expansion of the thirty-seventh column. No astronomical or calendrical theories have been put forward concerning the extra list columns. Space considerations in many tables mean that the list columns are often not present at all, and when they are, they are squashed into fewer than four columns. The motive for expansion therefore was surely not artistic.

An important requirement of Neugebauer and Parker's thirty-seven-column theory is that the column requires one more decan. A twelfth triangle decan must be postulated in order for the epagomenal column to record or model the sky in those five days in the same format as the rest of the table. The extra decan would occur only in the final cell of the epagomenal column. Demonstrating the existence of this decan, and therefore the validity of the epagomenal column as part of the main, functional body of the table (rather than as a non-functional member of the list columns) presents a problem which Neugebauer and Parker did not address.

The Composition of the Triangle in T and K

The triangle for the **K** family of tables is largely unknown, with no **K** tables containing any list columns nor indeed a single complete set of thirty-six ordinary columns. A fragmentary and questionable list of triangle decans can be reconstructed from the body of the tables and misplaced decans, with the possible aid of later astronomical diagrams: *štwy*, *sšpt*, *ipḏs*, *sbššn*, and *imy-ḫt 3ḫwy*. However, we have four sets of list columns from the **T** family (shown in Table 8), as well as some partial triangle areas, which provide a sufficient basis to discuss the epagomenal columns of *ṯm3t*-type tables.

The (reasonably secure) triangle for **T**, derived from the main body of **T** tables, is *smd rsy*, *smd mḥty*, *nṯr ḏ3 pt*, *rmn ḫry*, *ḫ3w*, *tpy-ʿ spd*, *imy-ḫt spd*, *3ḫwy*, *ḫ3w* (a second spelling indicating a different decan), *nṯr ḏ3 pt* (again), and *pḥwy s3bw*.

Neugebauer and Parker listed twelve triangle decans belonging to the *ṯm3t*-type decan lists, but their analysis of the final triangle decans is unconvincing as they had to postulate that the extra decan (which of course only occurs in the list columns) was always written out of order. In the main body, the eleventh triangle decan is *pḥwy s3bw*, yet Neugebauer and Parker labelled this the twelfth triangle decan and placed an otherwise unknown and difficult to read *s3bw* as eleventh decan. *S3bw* is only attested in coffins T6 and T8.

The writing of the extra decan *s3bw* in T6 shows it in the final position in the table, following *pḥwy s3bw*, whereas in T8, it is a tiny, intrusive label, written differently from all other decans in the table. This list finishes with *pḥwy s3bw* written twice.

Neugebauer and Parker explained the habitual appearance of *pḥwy s3bw* in eleventh place in four tables and T7's list column as a mistake, and preferred to interpret *s3bw* as the eleventh triangle decan, with the rather slim support of T8. Whether they insisted on transposing the order because the word *pḥwy* 'hindquarters' suggests it was not stated, but this must surely have been the reason. The evidence for a functioning epagomenal column is therefore far less conclusive than Neugebauer and Parker suggested.

Astronomically, the decans within the triangle should be in the area of the first eleven decans in the ordinary list. This is true for any theory of star tables, whether as clocks or almanacs, whether based on rising stars or other astronomical phenomena. However, the triangle decans in the *ṯm3t*-type tables certainly do not. They include *3ḫwy*, *ḫ3w*, and decans around *spd*. In other words, the **T** triangle seems to relate roughly to the third quarter of **T**, rather than the expected first third.

By the time the decans were used in astronomical ceilings (the New Kingdom onwards), we have seen that lists similar to **K** had become standard, and that a fragmentary triangle was usually present. This list contained up to seven triangle decans: *štwy*, *si3tw*, *nsrw*, *sšpt*, *nhs* or *ipḏs*, *sbššn*, and *nṯr w3š*. These decans are from the correct region, the first third, of the ordinary list.[29]

This implies that the astronomical diagram lists are closer to functioning decan lists. If the relationship between astronomical diagrams and **K** is considered to be valid, even if not strong, the **K** tables are, surprisingly, closer to functioning star tables than **T** tables.

Neugebauer and Parker saw the **T** tables as earlier than the **K** tables. Being 'earlier' is usually synonymous with being more genuine or 'uncontaminated' than later examples. Neugebauer and Parker reinforced their theory of the time-dependence of the decan lists by calculating[30] when the **T** triangle may have been constructed: 2780 BC (the early dynastic period). The confused state of the triangle in the surviving sources make even estimates of the date of the development very difficult to argue, even though the theory of an older triangle is credible.

T1	T6	T7	T8
35	35	35	23
36	36	1	1
1	1	2	2
2	2	3	3
3	3	4	4
4	5	5	5
5	5	6	23
6	6	7	24
7	7	8	8
8	8	9	9
9	9	10	10
10	10	13	11
11	11	14	12
12	12	16	26
13	13	17	14
14	14	18	16
15	15	20	17
16	16	23	18
18	17	24	20
17	20	26	C
18	22	27	D
20	23	28	26
21	24	30	27
22	26	31	28
23	27	32	30
24	28	33	27
25	30	34	32
26	31	35	33
27	32	A	34
28	33	C	35
29	34	D	A
30	35	J	F
31	29	K	A
32	A		J
33	B		L
34	C		K
35	D		K
2	E		
A	F		
B	L		
C	J		
D	K		
E			
F			
G			
J			
I			
K			

TABLE 8. Contents of the list and epagomenal columns of **T** class tables, the only tables to preserve these columns. The **T** class decan numbering system from Table 5 is used.

However, the existence of the triangle, the epagomenal label, and the tentative existence of a twelfth triangle decan makes it likely that Neugebauer and Parker were right to conclude that the epagomenal period was, at some point, fully accounted for in the diagonal star tables. This point alone raises these tables above the schematic, and places them as a high achievement in early astronomical activity.

If we review the place of diagonal star tables in the context of Egyptian astronomical activity we find that not only are they the only star tables to exhibit revisions, they are also the only ones to attempt to incorporate the entire length of the Egyptian civil year.

It has previously been argued[31] that the so-called 'transit star clock' is not a development of the diagonal star table, as Neugebauer and Parker claimed, but is instead its precursor. The 'transit star clock' text could describe an almanac of the events in the year of a decanal star and was formed with the intention of recording the movements of stars which had periods of invisibility of seventy days, similar to the 'ideal' decan Sirius and argued that seventy-day invisibility, with its funerary connections, provided a strong motive for the observations on which such a list must be based. Perhaps this motivation could also help explain the need for the greater detail displayed in the diagonal star table. If seventy days is an important factor, the 'transit' list would be inaccurate during the period when invisibility fell over the ignored epagomenal days. Following from the reasoning that the seventy-day period was the most important element, it can now be conjectured that this may be why the triangle was so carefully calculated or observed at some point in the history of the diagonal star table prior to the copying of the examples which still survive. Its subsequent descent into confusion is understandable both due to the complexity or imbalance of the area to the symmetrical Egyptian mind and the impact of uncomprehending copyists.

Hours and the Osireion Star Table

A schematic diagram of the surviving part of the Osireion star table is shown in Table 21. The third column is headed by a month name 'I Peret' and the label 'Hour Name'. In the six cells below, the names of the first six night 'hours' are listed.

Neugebauer and Parker showed[32] that this table is of the same nature as the diagonal star tables, but arranged differently. The fragment which survives contains eight decans.

The standard name for the tables, 'diagonal star clocks', indicates that the process of rising is thought to be strongly linked with time: the first star in the column performs its activity in (or at the beginning or end of) the first hour of the night, the second star in the second hour, and so on. However, a column of labels for the rows of the table is *not* present in the coffin-lid star tables. Row labels, analogous to the date row which labels the columns and allows us to associate these tables with the Egyptian year, would help enormously to identify the intended purpose of the tables.

Neugebauer and Parker's interpretation of the tables as 'clocks' was embedded throughout *Egyptian Astronomical Texts* Volume 1[33] where the word 'hour' was used on page one without discussion. Recently, researchers[34] have been questioning the assumption that these tables' primary use is timekeeping, and even whether the rows are related to hours at all. Perhaps the most obvious link to hours is that most tables contain twelve rows, however some tables have eight (K1 and K5), ten (T11), or even thirteen (K6) rows.

The Osireion star table represents a very important piece of evidence in this discussion, providing the only explicit link between diagonal star tables and time periods. In addition to the month name, the third cell in the first row of the surviving portion also contains the text *rn n wnwt* 'name of hour'. In the rest of that column, there are some labels which apparently give names to certain hours of the night, using the word *sp* 'period' in most cases and numbers two, three, and four in some of the cells: enough to show that time periods following sequentially from top to bottom are related to each of the 'rows' in the diagonal star table.

The Osireion star table tells us two things. First, that the structure of the table was known in the New Kingdom, emphasising the existence of source documents over and above the coffins which we know about. Second, that the understanding of the table in the New Kingdom was that it was related to periods of the night.

The Osireion table could provide the missing row labels for our First Intermediate Period coffin-based star tables if we are able to discount the possibility that the 'hours' relationship is a New Kingdom (or at least post-coffin-lid) development, perhaps borne out of a better understanding of the table after centuries of progress.

The fragmentary decan list in the Osireion table indicates that is closely related to some of the coffin-based diagonal star tables, but is separated by centuries from them in its execution. If a functioning, hour-based clock (with hours indicated in a column), was a New Kingdom development, it makes no sense for the old decans to be used. The date relationship would be, as we shall see, many decades wrong. Whoever added the hour names to this type of document did so on the understanding that hours were what the rows represented, and perhaps had always represented. It is therefore difficult to divorce this one column from the earlier tables, given the close relationship of the other three columns. Any theory which states that the rows or 'vertical axis' in the diagonal star tables does *not* represent time over the course of the night would also have to explain why this representation certainly *does* make that relationship. So far, such a theory has not been advanced.

Accepting that the rows of a diagonal star table are related to the hours of the night does not necessarily mean that the tables are clocks. Despite the New Kingdom use of the word 'hour' in connection with these tables, it is not the present author's current understanding that they should be considered 'clocks' in the sense of objects being designed and used primarily to tell the time.

Rising or Setting

Neugebauer and Parker also assumed from the first page of *Egyptian Astronomical Texts* Volume 1 that diagonal star tables were based on stars rising above the eastern horizon during the hours of darkness. They later stated that only by considering rising, could the date of the rising of Sirius within the tables be correlated with the historical dates of the painted star tables.[35] This assumption has recently been challenged by Leitz[36] who considered the *setting* of the stars to be the defining astronomical event of the tables.

Leitz based his setting theory on the position of Sirius in *ṯmꜣt* group tables. If settings are recorded, Sirius would set for the last time seventy days before I Akhet 1. Taking Sirius' period of invisibility to be seventy days,[37] Sirius would rise heliacally on I Akhet 1. The diagonal star table would therefore record an 'ideal' sky in an 'ideal' year. Although this is conceptually in keeping with Egyptian funerary texts, there are some problems with Leitz's argument.

First, Sirius is only positioned in this way in *ṯmꜣt* group sources. Leitz argued that one cannot know for certain that the columns in *knmt* group tables refer to the same decades as those in the *ṯmꜣt* group tables. However, this point has now been settled by the publication[38] of a fragment of an ordered *knmt* group table complete with date row showing that the first column is indeed headed by I Akhet First Decade.

Second, all *ṯmꜣt* group sources are missing two decans which Leitz agreed should be *before* Sirius, moving Sirius down the table and away from the I Akhet 1 position. Neugebauer and Parker considered[39] the possibility that Sirius was correctly positioned and that other decans should be moved to keep Sirius as the decan of the 12th hour in II Peret Final Decade. However, they suggested this amendment specifically to match the heliacal *rising* of Sirius with the historical date of the coffin lids—the opposite of what Leitz intended—so the two theories are not mutually supportive even if they require the same method of emendation of the tables.

The question of whether the tables were created based on settings or risings raises other points. As Leitz noted, setting horizons are different from rising horizons. That is, stars which rise together (forming a line on our celestial sphere) do not set together. The line of setting stars drawn on the celestial sphere is at a considerable angle to that of co-rising stars in the same region of the sky. The question of where to look for decans (on the east or west horizon) has a large impact on possible identifications of decans.

Leitz looked for setting decans and produced a scheme of decanal identifications. As always, these identifications rely heavily on the assumptions made concerning how the diagonal star tables were produced. The more assumptions that are made, the weaker the validity of any proposed identifications becomes. Yet raising new possibilities often sheds clearer light on old hypotheses as well.

Considering Leitz's setting theory, the difference between rising lines and setting lines in the Orion-Sirius region is striking. In **T**, there are seven Orion- or *sꜣḥ*-related decans (including *rmn ẖry*) between *ḫꜣw* and *spd*. In **K**, there are only four, with one being a possible triangle decan. In discussing problems with decan lists, previous researchers have theorised that human observational factors (such as tradition, eyesight, and personal preference) combined with practical observational factors (such as calibration methods and differences in location) could account for differences between decan lists such as splitting and combining decans. This type of explanation makes sense for the earlier parts of the decan lists **T** and **K** which agree to within a one-decan shift down to *ḫꜣw*, but cannot account for the severe discrepancy in the *ḫꜣw* to *spd* region.

Given that *ḫꜣw* and *spd* are each fixed objects (or in *ḫꜣw*'s case, perhaps a fixed group of objects: the name *ḫꜣw* means 'thousands' and could refer to an open cluster of stars like the Pleiades or Hyades), only a drastic explanation would solve the problem. Investigating rising and setting lines on a star map, one such theory presents itself: one decan list could have been based on rising stars, the other on setting stars. The lines of simultaneous settings cross the Orion region, as they do the rest of the sky, at a different angle from those of simultaneous risings. Fewer setting decans would be needed to span the region than rising decans.

This new theory of a pair of rising and setting lists, challenges another of Neugebauer and Parker's, and indeed all other researchers', assumptions. It has always been accepted that the two groups of diagonal star tables were a) objects of the same fundamental nature and b) separated by time with the *knmt* group being somewhat later than the *ṯmꜣt* group. The time separation is supposed to account for both the different starting decans and also the compositional differences between the **T** and **K** decan lists.

The paired rising and setting theory is unlikely, despite being attractive in the light of the Egyptian fondness for duality, due to the impact on the date of origin of the *knmt* group tables. Since its historical appearance sets a limit to the date of a decan list, **K** would have to be dated three hundred years *before* **T**. While it could be argued that greater antiquity could be responsible for the disorder of some *knmt* group tables, and that later monuments returned to the earlier and therefore more 'authentic' list (instead of continuing with the 'newer' list **T**), a theory of both rising and setting star tables is still a proposal that is as impossible to prove conclusively as a theory of solely setting diagonal star tables, or indeed, of solely rising tables. Only the discovery of further examples of diagonal star tables can elucidate the Orion area, perhaps refining the number of decans and their disposition in the complete decan lists.

However, the identification of a factor other than time as the basis of the two different classes is an important possibility which we will now discuss.

Revisions of the Diagonal Star Table

Neugebauer and Parker analysed both the date of the star tables[40] and the revisions which would have been necessary to keep the tables 'up to date', to counter the discrepancy between the solar year and the Egyptian civil year.[41] Their system was based on the fact that if the diagonal star tables record real astronomical events and could be used to generate 'time' during the night (the tables 'tell the time' not 'record the situation') the main body of the tables would be out of date in just forty years. If the epagomenal column was also to be kept up-to-date, the period between revisions would have to be as short as twenty years. So far, we have seen that star tables come in two broad 'families', strong evidence that at least one variant was produced. We have also noted that some star tables contain decan lists which are slight modifications of one of the major lists, perhaps suggesting some sort of interim adjustment. However, we have also seen that modification of the observing method could be responsible for major changes to the decan lists. Is it possible to distinguish between changes caused by revision over time and those brought about by other factors such as change in observing conditions?

If we assume that revisions were based solely on the passage of time, the effect on the diagonal star table can be predicted accurately. This allows us to compare the existing sources with the theoretical revisions.

To summarise Neugebauer and Parker's 'revision over time' theory, a revision which took place in twenty years after the date of the initial table would mean replacing each decan in the table (thirty-six ordinary decans numbered 1 to 36 and twelve triangle decans A to L) with a decan which had performed the defining action[42] five days *earlier* in the original table. The list of ordinary decans 1–36 would be replaced by twelve former triangle decans A to L plus new decans xii to xxxv. The triangle decans would be replaced by one new decan xxxvi plus original decans 1 to 11. The order of rising of the decans would be A, 1, B, 2, C, 3, D, 4, E, 5, F, 6, G, 7, H, 8, I, 9, J, 10, K, 11, L, 12, xii, 13, xiii, 14, xiv, and so on through to xxxiv, 35, xxxv, 36, xxxvi, A, 1, etc.

After another twenty civil years had passed, the table would once again be out of date. The next revision would have decan 36 in the first cell of the table. The main body of the table would be subject to another 'half-decan shift' with decan B being replaced with decan 1 and so on. The epagomenal column would have to be rewritten as 36, A, B, C, D, E, F, G, H, I, J, K.

This new table resembles the original star table of forty years before. The only difference is that a new decan, xxxvi,[43] now marks the first hour of the first decade of the year, and all other decans are shifted along one space diagonally. Decan L does not appear anywhere in the table, and decan K only appears once: as the final decan in the epagomenal column.

The change in lead decan is therefore the most noticeable indicator of the 'revision over time' theory. New lead decans ideally should not be drawn from the set of original decans. However, in the real sky, bright stars are not evenly distributed. Throughout the revision process, 'convenient' or well-known stars, or naturally darker regions of the sky might upset the rigid rules of revision. Regardless, the distance in star table cells between two decans should not vary by more than one cell from revision to revision if the observing method remains fixed.

We are now ready to look at the available star tables and compare them with this revision scheme. We have two classes of decan list, **T** and **K**, plus lists with minor variations as data. The first observation to make is that none of the sources currently known shows the major change in decans that would have occurred after a twenty-year revision. Second, the change in leading decan is consistent with the 'revision over time' theory. Some of the decans at the head of **K** are indeed different from those in **T**, indicating that intermediate decans could have been used. Third, that the 'distance' between decans is usually the same (to within one cell) except in the Orion area, the *ḫ3w* to *spd* distance mentioned above. Even though **T** and **K** are themselves only reconstructions, the discrepancy stretches the 'revision over time' mechanism to the limit. Fourth, that changes occur between the decan lists in areas where change is not required by the process, for example the splitting and joining of decans *wšt bk3t* and *spty ḫnwy*. In addition, comparison with the archaeological record shows that the relatively short period over which the coffins appear also lends little or no support to the 'revision over time' theory.

The process which resulted in the different classes of diagonal star tables seems to have been more variable and observationally-based than Neugebauer and Parker's system suggested. Factors other than the slippage of the Egyptian civil year against the observable sky are undoubtedly present. These factors could have included major changes using different events (rising, setting, or other actions of stars), or minor ones such as the location of the observations (perhaps indicating localised decan lists or competing traditions of lists, or perhaps variations in the horizon used), human and atmospheric observational factors, and unquantifiable pressures such as certain decans being aesthetically or symbolically favoured.

These remarks demonstrate that the assumption that the different decan lists representing a 'chronology' of observations is nowhere near as secure as Neugebauer and Parker's work indicated. The evidence does not support their deductions unequivocally. Once again, a hitherto accepted theory is shown to be only one of the available possibilities.

Summary of Conclusions

The following major points have been argued, described, or noted:

1) The corpus of diagonal star tables now numbers twenty-one sources.
2) Leitz's theory of a different date-basis for the later group of table has been disproved.
3) The inadequacy of previous attempts, primarily that of Neugebauer and Parker, to group or classify the tables has been demonstrated. Furthermore, classifications which include layout and epigraphy have been shown to be unlikely to produce appropriate and useful outcomes.
4) A system of classes based on decan lists has been proposed, currently having two main members **T** and **K**, and permitting variants such as **Ta** and **Ka**. Possible restorations of **T** and **K** have been presented and discussed. Notably, the order of decans *tpy-ʿ 3ḫwy*, *imy-ḫt 3ḫwy*, *3ḫwy* is deduced, contrary to Neugebauer and Parker, and the position of the *ṯm3t* decans at the head of tables has been questioned.
5) The relationship between 'later' or **K** class decan lists and lists in New Kingdom astronomical diagrams has been re-evaluated.
6) The importance of the existence of the triangle has been emphasised.
7) The independent existence of *s3bw* is queried and the weakness of the '37th column' theory has been demonstrated.

8) The existence of the hour label in the Osireion table must be explained if any theory of star tables which argues against the use of hours or time as the reason for rows in the tables is to be accepted.
9) A theory of the **T** and **K** list being a 'pair' rather than one being a re-working of the other has been introduced.
10) The revision history of the tables does not definitively indicate a system of revisions based solely on the passing of time, more complex or observationally-based scenarios such as that described in 9) above are consistent with the differences between decan lists.

Notes

1. Neugebauer and Parker (1960).
2. Full details of these thirteen sources will be found in Neugebauer and Parker (1960), pp. 4–23, together with references for each.
3. S1X, which Willems (*Chests of life* list of coffins) identifies as the coffin of *Ḥny*, (which is source 1 in *Egyptian Astronomical Texts* 3) is not a star clock, but an astronomical diagram. However, Lesko (*Index of the spells*) states S1X is an entirely different coffin belonging to *Ḏfꜣ.i-ḥꜥpy* whose texts have no references to decans.
4. Pogo (1936).
5. Pogo (1936) and Depuydt (1998), for example.
6. A written heading for this column is only preserved in two sources T7 and T8 and will be discussed presently.
7. Neugebauer and Parker (1960), p. 23.
8. Neugebauer and Parker (1960), p. 29.
9. Most researchers agree that the two identifications of *spdt* with Sirius and *sꜣḥ* with stars in the region of modern Orion are the only correspondences between ancient and modern names within the decans which are supportable. The precise extent, composition, and orientation of the ancient constellation *sꜣḥ* (that is, which stars formed various parts of the 'figure' of *sꜣḥ*) is, however, not known.
10. Leitz (1995).
11. Symons (2002b).
12. See section 'Rising or Setting' below.
13. This point has been demonstrated in the series of papers Locher (1983), Locher (1992), and Locher (1998). In the first paper, a diagram positioning each of the known sources (T1 to T8, T12, and K1 to K4) within a nest of intersecting lines. The lines separate factors such as date, provence, and epigraphical details, but omitting mention of the date row and the offering text. Two further tables (K5 and E1) were added in Locher (1992) by bending some of the lines. Locher abandoned the diagram in Locher (1998).
14. Kahl (1993).
15. As we will see, there is some uncertainty about whether the *ṯmꜣt* decans should head these tables or whether they are there by mistake. This group could arguably be designated the *wšꜣti* or **W** class instead. The labels **T** and **K** are based on the sources as they stand and have the benefit that they can be used to class new sources on first inspection.
16. Neugebauer and Parker (1960), pls. 26–29 listed seventy rising decans but two of these do not appear in any of the tables (*ṯs ꜥrḳ* and *sꜣḥ*), as Neugebauer and Parker themselves noted in their Additions and Corrections to Volume 1 which appeared in Neugebauer and Parker (1969), p. 272.
17. Neugebauer and Parker (1960), p. 23.
18. For example in the dramatic text from the *Book of Nut* see Neugebauer and Parker (1960), pls. 51–54.
19. Neugebauer and Parker estimated (Neugebauer and Parker (1960), 81) that the **T** class tables were created around 2150–2100 BC by noting that Sirius (assumed to be equivalent to *spd*) appeared in the 12th hour in the last decade of II Peret in T1 to T8. However, they noted (again Neugebauer and Parker (1960), p.

81) that this estimate was problematic due to the two omitted ordinary decans which may push the real date of 12th hour to the middle decade of III Peret and hence would make the tables eighty years younger. Given that the decan lists could not have been compiled after the coffins were painted, this suggests that *spd* is correctly placed and that the two missing decans should instead be inserted at some point before *spd* so that *ṯmꜣt ḥrt* and *ṯmꜣt ẖrt* would indeed be pushed off the beginning of the table (which would then be headed by *wšt bkꜣt*). This supports the discussion above regarding the placing of the *ṯmꜣt* decans. However, the date of Sirius marking the 12th hour of the night might not be the date of heliacal rise, as Neugebauer and Parker assumed. The 'decanal night' may have finished some time before the time when Sirius would first appear. This would mean that heliacal rise might take place one or even two decades before the star marked the 12th hour.

20. Such as the coffin of Heny (?IXth dynasty) described in Neugebauer and Parker (1969), pp. 8–10.
21. 'Typical' astronomical diagrams include the decans, the planets, and the circumpolar group. Other representations with astronomical content include ceilings based on funerary texts such as the Book of Nut and the Book of the Night.
22. The list of seven occurs in KV9 (Ramesses VI), see Neugebauer and Parker (1969), pp. 30–31.
23. The system was developed (Symons (1999) Section A) based on Neugebauer and Parker's twelve coffins plus the Osireion, with subsequent publications being incorporated without changing the structure behind the system.
24. Angular movement is 360/(23h 56m 4.09s) = 0.25 degrees per minute. Slippage per day (sidereal time versus solar time) is 24h - 23h 56m 4.09s = 3m 55.9s per day = 39m 19.1s over ten days. 39m 19.1s travelling at 0.25 degrees per minute results in an angular separation of 9.856 degrees.
25. Lacau (1906), pp. 101–128 and pl. 9, reproduced in Clagett (1995) fig. III.86.
26. Neugebauer and Parker (1960), p. 12.
27. Oriental Institute photograph P.26759/N.13354.
28. Neugebauer and Parker (1960), p. 1.
29. The locations of decans such as *ipḏs* and *sbššn* in relation to certain ordinary decans are known via the **T** group of tables and the decan list in the Book of Nut.
30. Neugebauer and Parker (1960) Chapter 3, section D.
31. Symons (2002a).
32. Neugebauer and Parker (1960), p. 32.
33. Neugebauer and Parker (1960), p. 1.
34. Leitz (1995), Depuydt (1998). In contrast, Clagett (1995) does not question the interpretation of Neugebauer and Parker.
35. Neugebauer and Parker (1960), p. 101.
36. Leitz (1995).
37. As recorded in the Dramatic Text from the *Book of Nut*.
38. Symons (2002b).
39. See footnote 19.
40. Neugebauer and Parker (1960), p. 31.
41. Neugebauer and Parker (1960), pp. 108–109.
42. For this examination the nature of the action, perhaps rising or setting, is not important as long as it is taken to be unchanging.
43. The process of updating the table at twently civil year intervals could in theory continue to keep this time-keeping method viable over the centuries, using a total of seventy-three decans: 1 to 36 (36 decans) plus A to L (12 decans) plus xii to xxxvi (25 decans) making 73 decans.

References

Borchardt, L., 1920, *Die Altaegyptische Zeitmessung* (Berlin and Leipzig: Walter de Gryter).

Clagett, M., 1995, *Ancient Egyptian Science Vol. 2: Calendars, Clocks, and Astronomy* (Philadelphia: American Philosophical Society).

Depuydt, L., 1998, "Ancient Egyptian Star Clocks and their Theory", *Bibliotheca Orientalis* 55, 5–44.
Eggebrecht, A., 1990, *Suche nach Unsterblichkeit* (Mainz: von Zabern).
——, 1993, *Antike Welt Pelizaeus-Museum Hildesheim: die aegyptische Sammlung* (Mainz: von Zabern).
Frankfort, H., 1933, *The Cenotaph of Seti I at Abydos* (London: Egypt Exploration Society).
Kahl, J., 1993 "Textkritische Bemerkungen zu den Diagonalsternuhren des Mittleren Reiches", *Studien zur Altaegyptischen Kultur* 20, 95–107.
Lacau, P., 1906, *Catalogue général des antiquités égyptiennes du Musée du Caire: Nos: 28087-28126: Sarcophages antérieurs au nouvel empire* Vol 2 (Cairo: IFAO).
Lange, H. O. and Neugebauer, O., 1940, *Papyrus Carlsberg No. 1. Ein hieratisch-demotischer kosmologischer Text*, Det Kongelige Danske Videnskabernes Selskab. Historisk-filologiske Skrifter, Bind I, Nr. 2 (Copenhagen: Ejnar Munsgaard).
Lapp, G., 1985, "Saerge des Mitteln Reiches aus der ehemaligen Sammlung Khashaba" *Aegyptologische Abhandlungen* 43, pl. 39.
Leitz, C., 1989, *Studien zur aegyptischen Astronomie* 1 (Wiesbaden: Harrasowitz).
——, 1995, *Altaegyptische Sternuhren*, Orientalia Lovaniensia analecta 62 (Leuven: Uitgeverij Peeters en Department Orientalistiek).
Lesko, L. H., 1979. *Index of the Spells on Egyptian Middle Kingdom coffins and related documents* (Berkeley: xxx).
Locher, K., 1983, "A Further Coffin Lid with a Diagonal Star Clock from the Egyptian Middle Kingdom", *Journal for the History of Astronomy* 14, 141–144.
——, 1992, "Two Further Coffin Lids with Diagonal Star Clocks from the Egyptian Middle Kingdom", *Journal for the History of Astronomy* 23, 201–207.
——, 1998, "Middle Kingdom Astronomical Coffin Lids: Extension of the Corpus from 12 to 17 Specimens since Neugebauer and Parker", in Eyre, C. (ed.), *Proceedings of the 7th International Conference of Egyptologists* (Leuven: Uitgeverij Peeters), 697–702.
Neugebauer, O., 1942, "The Origin of the Egyptian Calendar", *Journal of Near Eastern Studies* 1, 397–403.
——, and Parker, R. A., 1960, *Egyptian Astronomical Texts* Vol. 1 (Providence: Brown University Press).
——, and Parker, R. A., 1969, *Egyptian Astronomical Texts* Vol. 3 (Providence: Brown University Press).
Parker, R. A., 1950, *The Calendars of Ancient Egypt* (Chicago: University of Chicago Press).
———, (1974) "Ancient Egyptian Astronomy" in F. R. Hodson (ed.), *The Place of Astronomy in the Ancient World* (Oxford: Oxford University Press), 51–65.
Pogo, A., 1936, "Three Unpublished Calendars from Asyut" *Osiris* 1, 500–509.
Schaefer, B. E., 1985, "Predicting Heliacal Risings and Settings", *Sky and Telescope* 70, 261–263.
——, 1987, "Heliacal Rise Phenomena", *Journal for the History of Astronomy* 11, S19–33.
Symons, S., 1999, *Ancient Egyptian Astronomy: Timekeeping and Cosmography in the New Kingdom*, PhD Thesis, University of Leicester.
——, 2002a, "The 'Transit Star Clock' From the Book of Nut", in J. M. Steele and A. Imhausen (eds.), *Under One Sky: Astronomy and Mathematics in the Ancient Near East*, Alter Orient und Altes Testament 297 (Münster: Ugarit-Verlag), 429–446.
——, 2002b, "Two Fragments of Diagonal Star Clocks in the British Museum", *Journal for the History of Astronomy* 33, 257–260.
Willems, H., 1988, *Chests of Life: A Study of the Typology and Conceptual Development of Middle Kingdom Standard Class Coffins*, Mededelingen en Verhandelingen van het Vooraziatisch-Egyptisch Genootschap 'Ex Oriente Lux' 25 (Leiden; Ex Oriente Lux).

				IIII Shemu			III Shemu			II Shemu			I Shemu			IIII Peret			III Peret			II Peret			I Peret			IIII Akhet			III Akhet			II Akhet			I Akhet		
	27	14	1	L	M	F	L	M	F	L	M	F	L	M	F	L	M	F	L	M	F	L	M	F	L	M	F	L	M	F	L	M	F	L	M	F	L	M	F
A		13	36	36	35	34	33	32	31	30	29	28	27	26	24	23	22	21	20	18	17	16	15	14	13	12	11	10	9	8	7	6	5	4	3	2	1	36	35
B	26	14	1	A	36	35	34	33	32	31	30	29	28	27	26	24	23	22	21	20	18	17	16	15	14	13	12	11	10	9	8	7	6	5	4	3	2	1	36
C	27	15	2	B	A	36	35	34	33	32	31	30	29	28	27	26	24	23	22	21	20	18	17	16	15	14	13	12	11	10	9	8	7	6	5	4	3	2	1
D	28	16	3	C	B	A	36	35	34	33	32	31	30	29	28	27	26	24	23	22	21	20	18	17	16	15	14	13	12	11	10	9	8	7	6	5	4	3	2
E	29	18	4	D	C	B	A	36	35	34	33	32	31	30	29	28	27	26	24	23	22	21	20	18	17	16	15	14	13	12	11	10	9	8	7	6	5	4	3
F	30	17	5	E	D	C	B	A	36	35	34	33	32	31	30	29	28	27	26	24	23	22	21	20	18	17	16	15	14	13	12	11	10	9	8	7	6	5	4
G	31	18	6	F	E	D	C	B	A	36	35	34	33	32	31	30	29	28	27	26	24	23	22	21	20	18	17	16	15	14	13	12	11	10	9	8	7	6	5
J	32	20	7	G	F	E	D	C	B	A	36	35	34	33	32	31	30	29	28	27	26	24	23	22	21	20	18	17	16	15	14	13	12	11	10	9	8	7	6
I	33	21	8	H	G	F	E	D	C	B	A	36	35	34	33	32	31	30	29	28	27	26	24	23	22	21	20	18	17	16	15	14	13	12	11	10	9	8	7
K	34	22	9	I	H	G	F	E	D	C	B	A	36	35	34	33	32	31	30	29	28	27	26	24	23	22	21	20	18	17	16	15	14	13	12	11	10	9	8
to	35	23	10	J	I	H	G	F	E	D	C	B	A	36	35	34	33	32	31	30	29	28	27	26	24	23	22	21	20	18	17	16	15	14	13	12	11	10	9
tal	2	24	11	K	J	I	H	G	F	E	D	C	B	A	36	35	34	33	32	31	30	29	28	27	26	24	23	22	21	20	18	17	16	15	14	13	12	11	10

TABLE 9. Layout of T1 (S1C, 'Coffin 1'). Shaded cells indicate damage. In the date row, the decades are denoted by F for first decade, M for middle decade, and L for last decade. The decan numbers and letters are those in list **T**, shown in bold in Table 5.

III Shemu		II Shemu			I Shemu			IIII Peret			III Peret			II Peret			I Peret			IIII Akhet			III Akhet			II Akhet			I Akhet		
M	F	L	M	F	L	M	F	L	M	F	L	M	F	L	M	F	L	M	F	L	M	F	L	M	F	L	M	F	L	M	F
32	31	30	29	28	27	26	24	23	22	21	20	18	17	16	15	14	13	12	11	10	9	8	7	6	5	4	3	2	1	36	35
33	32	31	30	29	28	27	26	24	23	22	21	20	18	17	16	15	14	13	12	11	10	9	8	7	6	5	4	3	2	1	36
34	33	32	31	30	29	28	27	26	24	23	22	21	20	18	17	16	15	14	13	12	11	10	9	8	7	6	5	4	3	2	1
35	34	33	32	31	30	29	28	27	26	24	23	22	21	20	18	17	16	15	14	13	12	11	10	9	8	7	6	5	4	3	2
36	35	34	33	32	31	30	29	28	27	26	24	23	22	21	20	18	17	16	15	14	13	12	11	10	9	8	7	6	5	4	3
A	36	35	34	33	32	31	30	29	28	27	26	24	23	22	21	20	18	17	16	15	14	13	12	11	10	9	8	7	6	5	4
B	A	36	35	34	33	32	31	30	29	28	27	26	24	23	22	21	20	18	17	16	15	14	13	12	11	10	9	8	7	6	5
C	B	A	36	35	34	33	32	31	30	29	28	27	26	24	23	22	21	20	18	17	16	15	14	13	12	11	10	9	8	7	6
24	C	B	A	36	35	34	33	32	31	30	29	28	27	26	24	23	22	21	20	18	17	16	15	14	13	12	11	10	9	8	7
E	24	C	B	A	36	35	34	33	32	31	30	29	28	27	26	24	23	22	21	20	18	17	16	15	14	13	12	11	10	9	8
F	E	24	C	B	A	36	35	34	33	32	31	30	29	28	27	26	24	23	22	21	20	18	17	16	15	14	13	12	11	10	9
G	F	E	D	C	B	A	36	35	34	33	32	31	30	29	28	27	26	24	23	22	21	20	18	17	16	15	14	13	12	11	10

TABLE 10. Layout of T2 (S3C, 'Coffin 2'). In the date row, the decades are denoted by F for first decade, M for middle decade, and L for last decade. The decan numbers and letters are those in list **T**, shown in bold in Table 5.

20	17	18	15	14	13	12	11		10	9	8	7	6	5	4	3	2	1	36	35
21	20	17	18	15	14	13	12		11	10	9	8	7	6	5	4	3	2	1	36
22	21	20	17	18	15	14	13		12	11	10	9	8	7	6	5	4	3	2	1
22	22	21	20	17	18	15	14		13	12	11	10	9	8	7	6	5	4	3	2
23	22	22	21	18	17	16	15		14	13	12	11	10	9	8	7	6	5	4	3
24	23	22	22	20	18	17	16		15	14	13	12	11	10	9	8	7	6	5	4
26	24	23	22	21	20	18	17		16	15	14	13	12	11	10	9	8	7	6	5
27	26	24	23	22	21	20	18		17	16	15	14	13	12	11	10	9	8	7	6
28	27	26	24	23	22	21	20		18	17	16	15	14	13	12	11	10	9	8	7
29	28	27	26	24	23	22	21		20	18	17	16	15	14	13	12	11	10	9	8
30	29	28	27	26	24	23	22		21	20	18	17	16	15	14	13	12	11	10	9
31	30	29	28	27	26	24	23		22	21	20	18	17	16	15	14	13	12	11	10

TABLE 11. Layout of T3 (S6C, 'Coffin 3'). The decan numbers and letters are those in list **T**, shown in bold in Table 5.

	II Peret			I Peret		IIIIA		IIII Akhet			III Akhet			II Akhet			I Akhet	
L	M	F	L	M	F	L		M	F	L	M	F	L	M	F	L	M	F
16	14	15	12	11	10	9		9	8	7	6	5	4	3	2	1	36	35
18	16	14	15	12	11	10		10	9	8	7	6	5	4	3	2	1	36
17	18	16	14	15	12	11		11	10	9	8	7	6	5	4	3	2	1
20	17	18	16	14	15	12		12	11	10	9	8	7	6	5	4	3	2
21	20	17	18	16	14	15		15	12	11	10	9	8	7	6	5	4	3
23	21	20	17	18	16	14		14	15	12	11	10	9	8	7	6	5	4
24	23	21	20	17	18	16		16	14	14	12	11	10	9	8	7	6	5
26	24	23	21	20	17	18		17	16	15	14	12	11	10	9	8	7	6
27	26	24	23	21	20	17		18	17	16	15	14	12	11	10	9	8	7
28	27	26	24	23	21	20		20	18	17	16	15	14	12	11	10	9	8
29	28	27	26	24	23	21		21	20	18	17	16	15	14	12	11	10	9
30	29	28	27	26	24	23		22	21	20	18	17	16	15	14	12	11	10

TABLE 12. Layout of T4 (S1Tü, 'Coffin 4'). Shaded cells indicate damage. In the date row, the decades are denoted by F for first decade, M for middle decade, and L for last decade. The decan numbers and letters are those in list **T**, shown in bold in Table 5.

II Peret	I Peret				IIII Akhet				III Akhet			II Akhet			I Akhet	
F	L	M	F	L	M	F		L	M	F	L	M	F	L	M	F
14	13	12	11	10	9	8		6	6	6	4	3	2	1	36	35
15	14	13	12	11	10	9		8	5	5	5	4	3	2	1	36
16	15	14	13	12	11	10		9	8	7	6	5	4	3	2	1
17	16	15	14	13	12	11		10	9	8	7	6	5	4	3	2
18	17	16	15	14	13	12		11	10	9	8	7	6	5	4	3
20	18	17	16	15	14	13		12	11	10	9	8	7	6	5	4
21	20	18	17	16	15	14		13	12	11	10	9	8	7	6	5
22	21	20	18	17	16	15		14	13	12	11	10	9	8	7	6
23	22	21	20	18	17	16		15	14	13	12	11	10	9	8	7
24	23	22	21	20	18	17		16	15	14	13	12	11	10	9	8
26	24	23	22	21	20	18		17	16	15	14	13	12	11	10	9
27	26	24	23	22	21	20		18	17	16	15	14	13	12	11	10

TABLE 13. Layout of T5 (S2Chass, 'Coffin 5'). In the date row, the decades are denoted by F for first decade, M for middle decade, and L for last decade. The decan numbers and letters are those in list **T**, shown in bold in Table 5.

31	13	35	36	35	34	33	32	31	30	29	28	27	26	24	23	22	21	20	18	17		15	13	13	12	11	10	9	8	7	6	5	4	3	2	1	36	35
32	14	36	A	36	35	34	33	32	31	30	29	28	27	26	24	23	22	21	20	18		16	15	14	13	12	11	10	9	8	7	6	5	4	3	2	1	36
33	15	1	B	A	36	35	34	33	32	31	30	29	28	28	26	21	23	22	21	20		17	16	15	14	13	12	11	10	9	8	7	6	5	4	3	2	1
34	16	2	C	B	A	36	35	34	33	32	31	30	29	29	28	24	21	23	22	21		20	17	16	15	14	13	12	11	10	9	8	7	6	5	4	3	2
35	17 20	3	24	C	B	A	36	35	34	33	32	31	30	30	29	27	24	24	23	22		21	20	17	16	15	14	13	12	11	10	9	8	7	6	5	4	3
29	22	5	E	24	C	B	A	36	35	34	33	32	31	27	27	28	27	26	28	23		22	21	20	17	13	15	14	13	12	11	10	9	8	7	6	5	4
B A	23	5	F	E	24	C	B	A	36	35	34	33	32	31	30	29	28	27	26	24		23	22	18	20	17	16	15	14	13	12	11	10	9	8	7	6	5
C	24	6	G	F	E	24	C	B	A	36	35	34	33	32	31	30	29	28	27	26		17	23	21	21	18	17	16	15	14	13	12	11	10	9	8	7	6
D	26	7	H	G	F	D	24	C	B	A	36	35	34	33	32	31	30	29	28	27		24	17	22	22	20	18	17	16	15	14	13	12	11	10	9	8	7
E	27	8	I	H	G	E	E	24	C	B	A	36	35	34	33	32	31	30	29	28		26	24	23	23	21	20	18	17	16	15	14	13	12	11	10	9	8
F JL	28	9 10	J	I	H	F	F	E	24	C	B	A	36	35	34	33	32	31	30	29		27	26	24	17	22	21	20	18	17	16	15	14	13	12	11	10	9
K	30	12 11	K	J	I	G	G	F	E	D	C	B	A	36	35	34	33	32	31	30		16	27	26	24	17	22	21	20	18	17	16	15	14	13	12	11	10

TABLE 14. Layout of T6 (T3C, 'Coffin 6'). The decan numbers and letters are those in list **T**, shown in bold in Table 5.

			IIII Shemu			III Shemu			II Shemu			I Shemu			IIII Peret			III Peret				II Peret			I Peret			IIII Akhet			III Akhet			II Akhet			I Akhet		
5 days			L	M	F	L	M	F	L	M	F	L	M	F	L	M	F	L	M	F		L	M	F	L	M	F	L	M	F	L	M	F	L	M	F	L	M	F
31	13	35	36	35	34	33	32	31	30	29	28	27	26	24	23	22	21	20	18	17		16	15	14	13	12	11	10	9	8	7	6	5	4	3	2	1	36	35
32	14	1	A	36	35	34	33	32	31	30	29	28	27	26	24	23	22	21	20	18		17	16	15	14	13	12	11	10	9	8	7	6	5	4	3	2	1	36
33	16	2	B	A	36	35	34	33	32	31	30	29	28	27	26	24	23	22	21	20		20	17	16	15	14	13	12	11	10	9	8	7	6	5	4	3	2	1
34	17	3	C	B	A	36	35	34	33	32	31	30	29	28	27	26	24	23	22	21		21	20	17	16	15	14	13	12	11	10	9	8	7	6	5	4	3	2
35	18	4	D	C	B	A	36	35	34	33	32	31	30	29	28	27	26	24	23	22		22	21	20	17	16	15	14	13	12	11	10	9	8	7	6	5	4	3
A	20	5	E	D	C	B	A	36	35	34	33	32	31	30	29	28	27	26	24	23		23	22	21	20	17	16	15	14	13	12	11	10	9	8	7	6	5	4
C	23	6	F	E	D	C	B	A	36	35	34	33	32	31	30	29	28	27	26	24		18	23	22	21	20	17	16	15	14	13	12	11	10	9	8	7	6	5
D	24	7	G	F	E	D	C	B	A	36	35	34	33	32	31	30	29	28	27	26		24	18	23	22	21	20	17	16	15	14	13	12	11	10	9	8	7	6
J	26	8	H	G	F	E	D	C	B	A	36	35	34	33	32	31	30	29	28	27		26	24	18	23	22	21	20	17	16	15	14	13	12	11	10	9	8	7
K	27	9	I	H	G	F	E	D	C	B	A	36	35	34	33	32	31	30	29	28		27	26	24	18	23	22	21	20	17	16	15	14	13	12	11	10	9	8
	28	10	J	I	H	G	F	E	D	C	B	A	36	35	34	33	32	31	30	29		28	27	26	24	18	23	22	21	20	17	16	15	14	13	12	11	10	9
_	30		K	J	I	H	G	F	E	D	C	B	A	36	35	34	33	32	31	30		29	28	27	26	24	18	23	22	21	20	17	16	15	14	13	12	11	10

TABLE 15. Layout of T7 (G2T, 'Coffin 7'). In the date row, the decades are denoted by F for first decade, M for middle decade, and L for last decade. The decan numbers and letters are those in list **T**, shown in bold in Table 5.

5 days			IIII Shemu			III Shemu			II Shemu			I Shemu			IIII Peret			III Peret				II Peret			I Peret			IIII Akhet			III Akhet			II Akhet			I Akhet		
			L	M	F	L	M	F	L	M	F	L	M	F	L	M	F	L	M	F		L	M	F	L	M	F	L	M	F	L	M	F	L	M	F	L	M	F
27	26	23	31	13	35	36	35	34	33	32	31	30	29	28	24	22	21	20	18	17		16	15	14	13	12	11	10	9	8	7	6	5	4	3	2	1	36	35
32	14	1	A	36	35	34	33	32	31	30	29	28	27	26	24	23	22	21	20	18		17	16	15	14	13	12	11	10	9	8	7	6	5	4	3	2	1	36
33	16	2	B	A	36	35	34	33	32	31	30	29	28	27	26	24	23	22	21	20		20	17	16	15	14	13	12	11	10	9	8	7	6	5	4	3	2	1
34	17	3	C	B	A	36	35	34	33	32	31	30	29	28	27	26	24	23	22	21		21	20	17	16	15	14	13	12	11	10	9	8	7	6	5	4	3	2
35	18	4	24	C	36 35	B	A	36	35	33	32	31	30	29	28	27	26	24	23	22		22	21	20	17	16	15	14	13	12	11	10	9	8	7	6	5	4	3
A	20	5	E	D 24	36	35	B	A	36	35	33	32	31	30	29	28	27	26	24	23		23	22	21	20	17	16	15	14	13	12	11	10	9	8	7	6	5	4
F	C	23	6	F	E	D	C	B	A	36	35	34	33	32	31	30	28	27	26	24		18	23	23	22	21	20	17	16	15	14	13	12	11 10	9	8	7	6	5
A	D	24	7	F	E	D	C	B	A	36	35	34	33	32	31	30	29	28	27	26		24	18	23	22	21	20	17	16	15	14	13	12	11	10	9	8	7	6
LJ	26	8	H	G	F	E	D	C	B	A	36	35	34	33	32	31	30	29	28	27		26	24	18	23	22	21	20	17	16	15	14	13	12	11	10	9	8	7
K	27	9	I	H	G	F	E	D	C	B	A	36	35	34	33	32	31	30	29	28		27	26	24	18	23	22	21	20	17	16	15	14	13	12	11	10	9	8
	28	10	J	I	H	G	F	E	D	C	B	A	36	35	34	33	32	31	30	29		28	27	26	24	18	23	22	21	20	17	16	15	14	13	12	11	10	9
K	30	12 11	K	J	I	H	G	F	E	D	C	B	A	36	35	34	33	32	31	30		29	28	27	26	24	18	23	22	21	20	17	16	15	14	13	12	11	10

Table 16. Layout of T8 (A1C, 'Coffin 8'). In the date row, the decades are denoted by F for first decade, M for middle decade, and L for last decade. The decan numbers and letters are those in list **T**, shown in bold in Table 5.

I Shemu			IIII Peret			III Peret			II Peret				I Peret			IIII Akhet			III Akhet			II Akhet			I Akhet		
L	M	F	L	M	F	L	M	F	L	M	F		L	M	F	L	M	F	L	M	F	L	M	F	L	M	F
27	26	24	23	22	21	20	18	17	16	15	14		13	12	11	10	9	8	7	6	5	4	3	2	1	36	35
28	27	26	24	23	22	21	20	18	17	16	15		14	13	12	11	10	9	8	7	6	5	4	3	2	1	36
29	28	27	26	24	23	22	21	20	18	17	16		15	14	13	12	11	10	9	8	7	6	5	4	3	2	1
30	29	28	27	26	24	23	22	21	20	18	17		16	15	14	13	12	11	10	9	8	7	6	5	4	3	2
31	30	29	28	27	26	24	23	22	21	20	18		17	16	15	14	13	12	11	10	9	8	7	6	5	4	3
32	31	30	29	28	27	26	24	23	22	21	20		18	17	16	15	14	13	12	11	10	9	8	7	6	5	4
33	32	31	30	29	28	27	26	24	22	22	21		20	18	17	16	15	14	13	12	11	10	9	8	7	6	5
34	34	32	31	30	29	28	27	26	23	23	22		21	20	18	17	16	15	14	13	12	11	10	9	8	7	6
35	35	33	32	31	30	29	28	27	26	24	23		22	21	20	18	17	16	15	14	13	12	11	10	9	8	7
35	36	34	33	32	31	30	29	28	27	26	24		23	22	21	20	18	17	16	15	14	13	12	11	10	9	8
A	36	35	34	33	32	31	30	29	28	27	26		24	23	22	21	20	18	17	16	15	14	13	12	11	10	9
B	A	36	35	34	33	32	31	30	29	28	27		26	24	23	22	21	20	18	17	16	15	14	13	12	11	10

Table 17. Layout of T9 (S1Hil). Shaded cells indicate damage. In the date row, the decades are denoted by F for first decade, M for middle decade, and L for last decade. The decan numbers and letters are those in list **T**, shown in bold in Table 5.

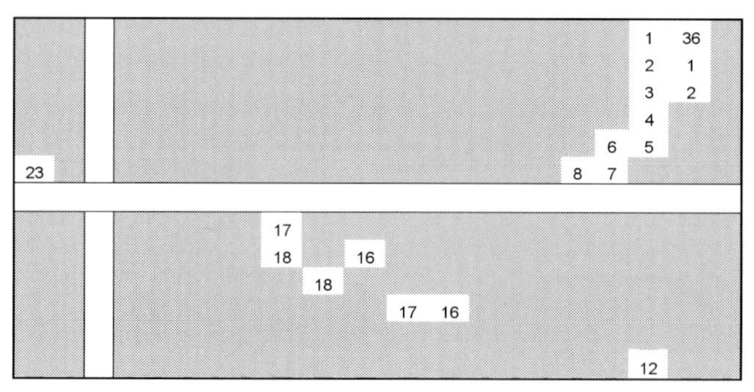

TABLE 18. Layout of T10 (S16C). Shaded cells indicate damage. The decan numbers and letters are those in list **T**, shown in bold in Table 5.

II Peret			I Peret			I Peret	IIII Akhet			III Akhet			II Akhet			I Akhet		
L	M	F	L	M		F	L	M	F	L	M	F	L	M	F	L	M	F
											6							
																5	4	
22	21		16			14				11	10	8			8			
23	22	21	17	16		14				12	11	9		10	9			
	23	22	21	17						13	12	10			10	9		
		23	22							14		11						
			23				16				14						10	9

TABLE 19. Layout of T11 (S2Hil). Shaded cells indicate damage. In the date row, the decades are denoted by F for first decade, M for middle decade, and L for last decade. The decan numbers and letters are those in list **T**, shown in bold in Table 5.

III Peret			II Peret			I Peret				IIII Akhet			III Akhet			II Akhet			I Akhet		
L	M	F	L	M	F	L	M	F		L	M	F	L	M	F	L	M	F	L	M	F
18	17	15	16	14	13	12	11	10		9	8	7	6	5	4	3	2	1b	1a	36	35
20	18	17	15	16	14	13	12a	11		10	9	8	7	6	5	4	3	2	1b	1a	36
21	20	18	17	15	16	14	13	12a		11	10	9	8	7	6	5	4	3	2	1b	1a
22	21	20	18	17	15	16	14	13		12a	11	10	9	8	7	6	5	4	3	2	1b
23	22	21	20	18	17	15	16	14		13	12a	11	10	9	8	7	6	5	4	3	2
24	23	22	21	20	18	17	15	16		14	13	12a	11	10	9	8	7	6	5	4	3
26	24	23	22	21	20	18	17	15		15	14	13	12a	11	10	9	8	7	6	5	4
27	26	24	23	22	21	20	18	17		16	15	14	13	12a	11	10	9	8	7	6	5
28	27	26	24	23	22	21	20	18		17	16	15	14	13	12a	11	10	9	8	7	6
29	28	27	26	24	23	22	21	20		18	17	16	15	14	13	12a	11	10	9	8	7
30	29	28	27	26	24	23	22	21		20	18	17	16	15	14	13	12a	11	10	9	8
31	30	29	28	27	26	24	23	22		21	20	18	17	16	15	14	13	12a	11	10	9

TABLE 20. Layout of T12 (S3P, 'Coffin 9'). Shaded cells indicate damage. In the date row, the decades are denoted by F for first decade, M for middle decade, and L for last decade. The decan numbers and letters are those in list **T**, shown in bold in Table 5.

	Third decade		Second decade	I Peret Name of Hour	First decade	
★	*ḥry-ib wiꜣ*	★	*ḥnwy*	*sp=s n ḫꜣw*	*sꜣpty*	★
★	*šsmw*	★	*ḥry-ib wiꜣ*	*sp 2=s m bkꜣt*	*ḥnwy*	★
★	*knmw*	★	*šsmw*	*sp 3=s m bkꜣt*	*ḥry-ib wiꜣ*	★
★	*tpy-ꜥ smd*	★	*knmw*	*sp 4=s m bkꜣt*	*šsmw*	★
★	*smd*	★	*tpy-ꜥ smd*	*?=s wšꜣw*	*knmw*	★
★	*srt*	★	*smd*	*sp=s nfr m wšꜣw*	*tpy-ꜥ smd*	★

I Peret		
Last	Middle	First
16	15b	15a
17	16	15b
18	17	16
19	18	17
20	19	18
21	20	19

TABLE 21. Layout of K0 (the star table in the sloping passage of the Osireion). The actual layout is shown on the left. On the right, the layout has been re-arranged into typical diagonal star table format. The decan numbers and letters are those in list **K**, shown in bold in Table 6.

	21	20	15b	16	15a	9	8	13		12	11	13	12	11	9	9	5	4	3	1	2	7	
23	22	21	20	17	15b	15a	9	8		13	12	11	13	12	11	9	9	5	4	3	1	2	7
24	23	22	21	20	17	15a	15b	15a		8	13	12	11	13	12	11	9	9 8	5	4	3	1	2
	24		22	21		18	17	15b		9	8	11	13	8	9	9	11	9	9	5		4	1
	25	24	24	20	24	3	17	30		14	15a	12	8	9	11	15a	8	9	8	8	5	4	
	25	25	25	21	25	24	18	29		15b	15a	14	9	11	15a	15b	12	11	9	9	6	5	
28	29	26	26	22	E	E	20	20		15a	15b	15a	11	8	15b	16	13	12	11	8	7	6	
	30	29	27	23	25	25	29	29		16	16	15b	12	9	16	8	15b	13	12	9	8	7	

TABLE 22. Layout of K1 (S9C, 'Coffin 10'). Shaded cells indicate damage. The decan numbers and letters are those in list **K**, shown in bold in Table 6.

24	23	22	21	20		17	16		5	5	5	10	13	12	11	6	8	8	7	6	10	3	2	1
23	22	21	20	16		16	14	5	4	4	4	10	13	12	1	8	2	3	3	10	9	2	1	2
	21	20	16	4		14	15	4	10	10	10	8	2	2	8	2	7	10	10	9	8	1	4	4
	20	16	4	5		15	16	10	8	2	8	13	12	11	2	7	6	9	9	8	2	4	5	5
20	16	4	5	10		16	17	8	13	2	13	12	11	10	7	6	9	8	8	2	1	6	6	6
16	4	5	10	15		17		13	2	13	12	11	10	6	6	9	2	2	2	1	2	7	7	7
	16	20	D	36		4	8	10	6	6	18	18	19	21	22	2	1	25	26	27	28	4	5	31
16	20	17	36	35		8	10	6	7	7	19	19	20	22	2	1	25	26	27	28	4	5	31	29
	17	D	35	A		10	6	7	2	2	20	20	21	2	1	25	27	27	28	4	5	31	29	32
	D	36	A	B		6	7	2	18	19	21	21	22	1	25	27	28	28	4	5	31	29	32	33
	36	35	B	32		7	2	18	19		2	2	2	25	27	28	2	2	5	31	29	32	33	34
	35	32	32	8		2	18	19	20		22	22	1	27	28	2	6	6	31	29	32	33	34	29

TABLE 23. Layout of K2 (S5C, 'Coffin 11'). Shaded cells indicate damage. The decan numbers and letters are those in list **K**, shown in bold in Table 6.

18	17	16	15	14	13	12		11	9	8	7	6	5	4	3	2	1
19	18	17	16	15	14	13		12	11	9	8	7	6	5	4	3	2
20	19	18	17	16	15	14		13	12	11	9	8	7	6	5	4	3
21	20	19	18	17	16	15		14	13	12	11	9	8	7	6	5	4
22	21	20	19	18	17	16		15	14	13	12	11	9	8	7	6	5
23	22	21	20	19	18	17		16	15	14	13	12	11	9	8	7	6
24	23	22	21	20	19	18		17	16	15	14	13	12	11	9	8	7
26	24	23	22	21	20	19		18	17	16	15	14	13	12	11	9	8
27	26	24	23	22	21	20		19	18	17	16	15	14	13	12	11	9
28	27	26	24	23	22	21		20	19	18	17	16	15	14	13	12	11
29	28	27	26	24	23	22		21	20	19	18	17	16	15	14	13	12
30	29	28	27	26	24	23		22	21	20	19	18	17	16	15	14	13

TABLE 24. Layout of K3 (S11C, 'Coffin 12'). The decan numbers and letters are those in list **K**, shown in bold in Table 6.

19	18	17	16	15	14	13	12		11	10	9	8	7	6	5	4	3		
20	19	18	17	16	15	14	13		12	11	10	9	8	7	6	5	4	3	
21	20	19	18	17	16	15	14		13	12	11	10	9	8	7	6	5	4	3
22	21	20	19	18	17	16	15		14	13	12	11	10	9	8	7	6	5	4
23	22	21	20	19	18	17	16		15	14	13	12	11	10	9	8	7	6	5
24	23	22	21	20	19	18	17		16	15	14	13	12	11	10	9	8	7	6
25	24	23	22	21	20	19	18		17	16	15	14	13	12	11	10	9	8	7
28	25	24	23	22	21	20	19		18	17	16	15	14	13	12	11	10	9	8
27	27	26	24	23	22	21	20		19	18	17	16	15	14	13	12	11	10	9
29	28	28	25	24	23	22	21		20	19	18	17	16	15	14	13	12	11	10
30	29	28	28	25	24	23	22		21	20	19	18	17	16	15	14	13	12	11
31	30	29	28	28	25	24	23		22	21	20	19	18	17	16	15	14	13	12

TABLE 25. Layout of K4 (S#C). Shaded cells indicate damage. The decan numbers and letters are those in list **K**, shown in bold in Table 6.

13	12	11	10	9		8	7	6	5	4	3	2
14	13	12	11	10		9	8	7	6	5	4	3
15	14	13	12	11		10	9	8	7	6	5	4
16	15	14	13	12		11	10	9	8	7	6	5
17	16	15	14	13		12	11	10	9	8	7	6
18	17	16	15	14		13	12	11	10	9	8	7
20	18	17	16	15		14	13	12	11	10	9	8
21	20	18	17	16		15	14	13	12	11	10	9

TABLE 26. Layout of K5 (X2Bas). Shaded cells indicate damage. The decan numbers and letters are those in list **K**, shown in bold in Table 6.

I Akhet First
1
2
3
4
5
6
7
8
9
10
11
12
13

TABLE 27. Layout of K6 (EA47605). The decan numbers and letters are those in list **K**, shown in bold in Table 6.

1
2
3

TABLE 28. Layout of K7 (no number). The decan numbers and letters are those in list **K**, shown in bold in Table 6.

Calendars and Years in Ancient Egypt: The Soundness of Egyptian and West Asian Chronology in 1500–500 BC and the Consistency of the Egyptian 365-Day Wandering Year

Leo Depuydt

My fellow historians of antiquity! The state of traditional ancient chronology is *strong*. There always have been episodes of doubt. There always will be need for introspection and verification. But the standard chronology found in all the textbooks will prevail!

What follows is an exercise in introspection of the chronological kind. The focus is on a fundamental tenet of the reigning chronology of ancient Egypt. By this tenet, the Egyptian calendar shifted or wandered with absolute consistency at a rate of about one day in four years in relation to the seasons throughout Egyptian history, for more than 3000 years, from the early third millennium BC onward. Back to about 500 BC, there is hardly a doubt about the veracity of the absolute consistency of the Egyptian calendar's wandering. The present concern is with the millennium that immediately precedes 500 BC, namely the time-period 1500–500 BC. The objective of this paper is to argue in favour of the veracity of absolute consistency for that time-period as well.

The following logic is fundamental to the organization of this paper's argument. The standard ancient chronology for the time-period 1500–500 BC *implies* absolute consistency of the Egyptian calendar's wandering. At the same time, the chronology of 1500–500 BC seems secure for reasons *independent of* the consistency of the Egyptian year's wandering. Therefore, logically speaking, to the extent that the chronology of 1500–500 BC can be considered secure, the absolute consistency of the Egyptian year's wandering in that same time-period is also secure by inference, because the former logically implies the latter. A principal concern of the present paper is therefore the theoretical model that has produced the chronology of 1500–500 BC. The structure of the chronology of that time-period is analysed in section 5 of this paper.

In addition, the time-period 1500–500 BC will be placed in a larger context by treating milestones in the history of the Egyptian 365-day calendar and in the history of the study of that calendar from before and after 1500–500 BC. The present treatment of these milestones may serve as a sort of prolegomena to a more comprehensive survey of the history of the Egyptian 365-day calendar. Four milestones will precede the analysis of the chronology of 1500–500 BC in sections 1, 2, 3, and 4. One milestone will follow in section 6.

As for the order in which the milestones will be treated, the following organizational principle rules. History is a story. Stories are best told *forward* in time. As the backbone of history, chronology is all about time. Yet, unlike history proper, chronology is best told *backward*, from the present back into the past. By moving backward, one only follows the sound scientific principle of proceeding from what is more certain to what is less certain. Accordingly, the following six milestones will be treated below in reverse chronological order.

(1) FIRST MILESTONE (AD 1820s)
Champollion discovers the seasonal month names in hieroglyphic sources

(2) SECOND MILESTONE (AD 1806)
Ideler first fully evaluates the chronological data in Ptolemy

(3) THIRD MILESTONE (AD 1578)
Crusius puts Ptolemy's era of Nabonassar in order

(4) FOURTH MILESTONE (Fifth Century BC)
Babylonian and Egyptian dates as a rule match in Aramaic double dates

(5) FIFTH MILESTONE (1500–500 BC)
Why is Egyptian and West Asian chronology in 1500–500 BC sound?
The Amarna and Assyrian Connections

(6) SIXTH MILESTONE (Early Third Millennium BC)
The origin of the Egyptian 365-day calendar

* * *

0. Preface

0.1. The Limits of Induction

Intellectual disciplines operate by theoretical models. Only the deductive models of mathematics and mathematical logic are absolutely certain. And that does not even include the axioms on which chains of deductive steps are built. All other models are inductive. Inductive models consist of patterns of regularity emerging from repeated observation of facts. Facts are typically instances of contact between the senses and the world outside the senses. Inductive models may come quite near to certainty. But they are never entirely divested of probability. And that includes the laws of nature, which are inductive.

There is always room for skeptics to doubt inductive models. To be successful, inductive models need to earn and win the disinterested, unsolicited, and unforced consensus of a number of independent observers. Absolute certainty does not exist. Universal consensus is not possible. But there is winning and there is losing. A winning model allows a majority of observers to stand their ground confidently over the decades and over the centuries against the inevitable skepticism. One measure of a model's success is its ability to dominate the textbooks. Then again, inductive models are never permanently established. The need for confronting certain kinds of persistent skepticism never vanishes. Accordingly, creationism will always pose a challenge to evolution theory. An inductive model is a dynamic process. There is a never-ending exchange between its advocates and its skeptics. But in this process, certain models remain on top while rival theories are relegated to the status of also-rans.

Unforced and unsolicited consensus has a peculiar way of overpowering skepticism, however loudly proclaimed. The power of consensus has much to do with how consensus may be quantified. An agreement between two (2) observers may be counted as one (1) unit of consensus. Marginal models are often publicized so loudly that they seem to drown out the standard model. But in spite of the loudness, no two observers can be found to agree on the same marginal model. The consensus quote of such marginal models is therefore zero. And zero is still infinitely larger than a consensus of just a handful of keen observers. Disagreeing is not too difficult. There is always room for skepticism. But earning the consensus of other observers regarding the exact formulation of an alternative chronological model is less easy.

Our modern calendar, instituted by Caesar in 45 BC, provides a kind of continuity down to the present time that makes chronology much less of a concern to AD historians than to BC historians.[1] Handbooks of ancient history all agree on the outline of BC chronology. This universal agreement sharply contrasts with an almost equally universal lack of awareness of how this outline was obtained.

The standard ancient chronology is a product of an inductive chronological model. This model is an intellectual structure. The structure consists of various components that are built upon one another in a logical sequence. Some components are logically prior to other components and therefore more fundamental.

No inductive model is above constant verification. Chronological models, being inductive as they are, are no exception. Reluctance to verify the standard model of ancient chronology may engender suspicion that all is not well with ancient chronology. Such suspicion in fact keeps spawning wholesale rejections of the standard chronology in the margins of the field of history.

Lacking proper understanding of the standard model's foundation and structure, historians cannot defend themselves against refutations of the model. Whether the standard model is true or false is at first irrelevant. What matters at the outset is that the model is very real in the following fundamental respect. It is responsible for the fact that all the history books agree when Ramses II reigned. The model is the hidden engine that drives all the dates in all the textbooks. This engine's internal workings are grasped to their full extent by hardly anyone. The discipline of ancient studies would be stronger if more people understood how the model works.

Towards this aim, I have portrayed the model of ancient chronology back to about 500 BC in four recent contributions, each endowed with its own distinct focus. The first contribution is an article entitled "Ancient Chronology's Alpha and Egyptian Chronology's Debt to Babylon", to appear in a Festschrift published by Eisenbrauns. Its focus is specifically to outline the model. The second contribution is a shorter and popular version of this first contribution, addressed to a wider audience, and published under the title "How to Date a Pharaoh" in the issue of July/August 2005 of the magazine *Archaeology Odyssey*, at pp. 27–33. The third contribution consists of two chapters entitled "Foundations of Day-exact Chronology" and "Saite and Persian Egypt, 664–332 BC (Dynasties 26–31, Psammetichus I to Alexander's Conquest of Egypt)" in the new handbook of Egyptian chronology for the series Handbuch der Orientalistik (now just out as Depuydt (2006a) and (2006b)). The objective of these two chapters is to survey the chronology of 664–332 BC. The fourth contribution is an article entitled "The Shifting Foundation of Ancient Chronology" in the acts of a session held at the Ninth Annual Meeting of the European Association of Archaeologists in 2003 in St. Petersburg.[2] This contribution puts more than four centuries of modern study of chronology, beginning in the early seventeenth century AD, in perspective by focusing on how the fundaments on which the entire structure of ancient chronology rests are shifting and will be shifting in the years ahead from Ptolemy's Royal Canon to the Babylonian astronomical diaries and related sources.

The aim of these four contributions was to make the chronology back to about 500 BC more transparent and accessible. The aim of the present paper is to move this boundary back to about 1500 BC, in the service of the ulterior aim of defending the absolute consistency of the Egyptian wandering year.

0.2. General Characteristics of the Egyptian 365-Day Calendar

0.2.1. Definition of the Consistency of the 365-Day Calendar's Wandering

A deep concern for the veracity of the foundations of ancient chronology has inspired the organization of the session on "Calendars and Years" held at the July 2005 meeting of the biannual Notre Dame Workshop on the History of Astronomy. Accordingly, the focus of this paper is on the most fundamental components of the standard model of ancient chronology, the ones on which everything else rests.

In Egyptian chronology, few assumptions are as fundamental as that of the basic mechanism of the 365-day calendar. This mechanism is eminently simple. Egyptian years consist of 12 months of 30 days plus five added days, for a fixed total of 365 days (12 × 30 + 5). There were no leap years of 366 days. Because the year of the seasons is about a quarter day longer than 365 days, the Egyptian year shifts in relation to the seasons at a rate of about one day in four years. In relation to our modern calendar as extended artificially backward into the past, this shift is *exactly* one day in four years. If New Year's Day falls on 31 December, it will fall on 30 December four years later, on 29 December after another four years, and so on. New year moves backward from winter into fall and then into summer and spring, to return to the same day in winter after 1460 years. It is generally assumed that the Egyptian year followed this shifting or wandering course for more than three thousand years. Very early in the first millennium AD, this shifting year ceased being used in the calendar of daily life. Astronomers retained it for many centuries more because it is so simple.

No assumption is more essential to the current model of ancient Egyptian chronology than the following. It is that the Egyptians never tinkered with the regular backward shift of the Egyptian year in relation to the seasons for more than three millennia, from the early third millennium BC onward. They were never tempted to add days to or subtract days from the Egyptian 365-day calendar. Accordingly, not a single year ever had more or fewer than 365 days. The calendar therefore wandered consistently at the same pace in relation to the seasons. Current Egyptian chronology stands or falls with the assumption of such absolute consistency on the part of the calendar's wandering. If the wandering was not consistent, then the standard Egyptian chronology is wrong. Is it justified to assume consistency?

Proving consistency *directly* is impossible. Evidence regarding the length of each single Egyptian year for a period of more than 3000 years would be required. No such detailed records are available or have survived. Is the standard chronology of ancient Egypt then a giant with clay feet? Can Egyptian chronology be falsified in its totality merely by pulling out the single card of the wandering year's consistency at the bottom so that the whole cardhouse comes tumbling down?

In the absence of detailed records, the need is for *indirect* evidence. One type of indirect evidence is as follows. It applies back to about 500 BC. The evidence involves the following reasoning. The Egyptian year wanders in relation to other calendars and to the seasons. If—assuming consistency—the year is where it is supposed to be at two points in time separated from one another by an interval of time, it may be inferred where the year was everywhere in between. For example, if New Year's Day falls on 31 December in one year, one expects it to fall eight days earlier after 32 years. The year moves backward by one day in four years (4 × 8 = 32). If the sources somehow confirm 23 December as the date of New Year's Day, then it can be inferred when the Egyptian New Year's Day fell in each of the intervening years. If the calendar had been altered once in that interval, a second alteration would be needed to cancel the first and return the calendar to its regular course. It is improbable that two unattested calendrical adjustments could exactly cancel one another.

0.2.2. General Considerations Favouring the Consistency of the 365-Day Calendar's Wandering

Certain considerations favouring the assumption of the wandering year's consistency do not rise to the level of proof. Nor is it clear whether they apply to the period before 500 BC. Some of these considerations are as follows. First, why would anyone have changed the calendar by adding or dropping days? The only reasonable motive is to make the year stop wandering in relation to the seasons. But we know that it wandered throughout Egyptian history. That is because the known Sothic dates shift over the centuries. Sothic dates are dates of the first morning rising of Sirius in the Egyptian calendar. The rising stays roughly the same in relation to the solar year. If the Egyptian year wandered, then Sothic dates ought to shift over time, as they in fact do.

A second consideration is testimonies that the Egyptian year wandered with absolute regularity. These testimonies are cited in handbooks of chronology. The first testimony is in Greek by Geminus, a respected astronomer who lived in the first century BC. Geminus also cites the testimony of Eratosthenes of the third century BC about the wandering motion of the Egyptian calendar. The second testimony is by Nigidius Figulus, a learned contemporary of Varro. Varro consulted him about the chronology of Rome's founding. Nigidius Figulus transmits a tradition that Pharaohs swore never to intercalate days. These first two testimonies are in Greek or Latin and are late. One might be inclined to question what Greek and Roman authors knew about Egypt. The third testimony, however, is in Egyptian. It is found in the Canopus decree of 238 BC. This decree is an official text about an agreement between the highest priestly authorities and the king. Stelae inscribed with copies of the decree were erected all over Egypt. The Canopus Decree is in three versions in three languages, two of them being stages of Egyptian. One would expect the text to have been drawn up with care. The authors belonged to the priest-scribe caste that preserved hieroglyphic and Pharaonic religion and culture. The text speaks about the "displacement in appearances by the star by one day every four years . . ., which happened in the past" (Tanis version, lines 40–41). The star is Sirius. Its displacement in the calendar of one day in four years is presented as the rule. This displacement is what one would expect if the 365-day calendar wandered regularly at all times. There are many indications that Egyptian priest-scribes carefully preserved hieroglyphic culture and jealously guarded it over many centuries. Still, could an irregularity in the calendar happening many centuries before the third century BC somehow have passed into oblivion?

A fourth consideration is freedom of contradiction. No surviving evidence blatantly contradicts the reigning chronology. No widely recognized major obstacles are in its way. All new evidence that has emerged has easily fitted into the reigning model. Nor does there exist anything close to a comprehensive alternative model that coherently and systematically accounts for all the evidence. There are only isolated efforts at maximally exploiting the inevitable deficit from absolute certainty exhibited by all inductive theories. The display of skepticism is ultimately not sufficient or satisfactory. Skepticism requires the support of an alternative model that comprehensively explains all the surviving evidence while avoiding contradiction.

0.2.3. Relation of the Egyptian 365-Day Calendar to Other Egyptian Calendars

The 365-day calendar does not follow the lunar cycle. But obviously, the fixed artificial length of its months, namely 30 days, is derived from the lunar cycle. As regards lunar time-reckoning, Egypt is unique among ancient nations. In all other nations, the lunar calendar was dominant. The Babylonians, the Greeks, the Hebrews, and so on, all regulated daily life by the moon. So did Rome until the fifth century BC, when it switched to a complex calendar that exhibits peculiar vestiges of lunar time-reckoning.

This does not mean that there was no lunar time-reckoning at all in Egypt. Lunar time-reckoning was used marginally in religious life, for instance, in order to regulate the service in the temple. Egypt therefore had two calendars, the dominant 365-day calendar, often called the civil calendar, and the somewhat marginal lunar calendar.

At some point, these two calendars became attached to one another. One might call these two calendars winding forward in time in conjoined fashion a double helix calendar. Egyptian calendrics has a reputation for being complex. Many of the key problems pertain to the structure of this double calendar.

Egyptian calendrics can be telescoped, in my opinion, into the following 17-word formula: *A spawned I plus one seeming instance of II and, once A had happened, B spawned II.* The formula owes its absolute logical anteriority to the fact that it presupposes knowledge of everything that is essential about the fundamental structure of Egyptian calendars. A and B are postulated actions by ancient calendar-makers. I and II are phenomena securely attested in the sources. Definitions of I, II, A, and B are as follows. Empirical phenomenon I (the "Brugsch" phenomenon): The last month can be named as if it were the first, namely *wp-rnpt* or *mswt-rʿ*, which are also names of New Year's Day, the quintessential beginning. Empirical phenomenon II (the "Gardiner" phenomenon): A feast can sometimes be celebrated on Day 1 of the civil month that immediately follows the civil month bearing the same name as the feast—as if a feast called January were celebrated on 1 February. Action A: Month names are transferred from the lunar calendar to the civil calendar. Action B: Feast days are transferred from the lunar calendar to the civil calendar. Egypt had a dominant 365-day civil calendar and a marginal lunar calendar. There is no firm evidence for any other calendars. The aforementioned formula concerns the relation between these two calendars.[3]

0.2.4. Progression of the Egyptian 365-Day Calendar in Relation to the Julian Year

The Egyptian year is just under a quarter day shorter than the year of the seasons of about 365.2422 days. The Egyptian year therefore shifts or wanders in relation to the seasons at a rate of just under one day in four years. This wandering motion can be exemplified by means of the position of the Egyptian wandering year in relation to the Julian new year. The Julian year is on average exactly 365¼ days long. The Egyptian year therefore shifts exactly one day in four years in relation to the Julian year. The following table lists the time intervals covered by 12 successive Egyptian 365-day years in the Julian calendar. The year dates are BC. The Julian calendar, including years BC, did not yet exist at the time. The calendar is artificially extended into the past by a modern convention first introduced in the seventeenth century AD. The 365-day years in italics contain an intercalary February 29.

5 September 53	–	4 September 52	
5 September 52	–	4 September 51	
5 September 51	–	4 September 50	
5 September 50	–	*3 September 49*	[contains 29 February 49]
4 September 49	–	3 September 48	
4 September 48	–	3 September 47	
4 September 47	–	3 September 46	
4 September 46	–	*2 September 45*	[contains 29 February 45]
3 September 45	–	2 September 44	
3 September 44	–	2 September 43	
3 September 43	–	2 September 42	
3 September 42	–	*1 September 41*	[contains 29 February 41]

As one can see, the Egyptian New Year's Day, which is the date to the left, shifts backward in relation to the artificial Julian calendar. The one-day shifts occur after every 29 February, when the Julian calendar had one day more and the Egyptian calendar did not.

A different manner of illustrating the wandering motion of the Egyptian 365-day year is to list the Julian dates of the Egyptian New Year's Day at certain intervals. As one can see, each Julian date is valid for four years. The Julian dates 1 January, backwards to 1 October, backwards to 1 July, and so on, have been chosen. To the left are listed the sets of four Julian years in which the Egyptian New Year's Day falls on the Julian day and month date to the right.

1984–1981 BC:	1 January
1617–1614 BC:	1 October
1249–1246 BC:	1 July
885–882 BC:	1 April
524–521 BC:	1 January
157–154 BC:	1 October
AD 212–215:	1 July

In the sequence 1984, 1617, 1249, 885, 524, 157, and 212, there are shifts from even numbers to uneven numbers and back to even numbers. The shift from even to uneven happens at the transition from 1 January back to 31 December. The shift from uneven to even happens at the transition from 1 March to 29 February to 28 February.

These shifts have everything to do with the fact that 1460 Julian years contain exactly the same number of days as 1461 Egyptian years, namely 533,265 days. Indeed, 1461 (Egyptian years) × 365 = 1095 (Julian normal years) × 365 + 365 (Julian leap years) × 366 = 533,265 days. In that period of 533,265 days, there are exactly 365 Julian leap years of 366 days. Every fourth Egyptian year will therefore encompass a Julian 29 February. After 1460 Egyptian years, the 365 instances of Julian 29 February will have been exhausted, as it were. But there are 1461 Egyptian years in the period at hand. There will therefore be one sequence of *five* Egyptian years containing just one 29 February. That sequence occurs when the Egyptian New Year's Day shifts from 1 March to 29 February to 28 February. The details of this transition are seen in the following list of 14 successive 365-day years.

3 March 766	–	*1 March 766*	[contains 29 February 765]
2 March 765	–	1 March 764	
2 March 764	–	1 March 763	
2 March 763	–	1 March 762	
2 March 762	–	*29 February 761*	[contains 29 February 761]
1 March 761	–	28 February 760	
1 March 760	–	28 February 759	
1 March 759	–	28 February 758	
1 March 758	–	28 February 757	
29 February 757	–	*27 February 756*	[contains 29 February 757]
28 February 756	–	27 February 755	
28 February 755	–	27 February 754	
28 February 754	–	27 February 753	
28 February 753	–	*26 February 752*	[contains 29 February 753]

Between 29 February 761 and 29 February 757, not counting these two days, there are exactly 1460 days, which amount to four full Egyptian 365-day years, beginning with the New Year's Day of the first of the four years. In the early part of the above list, Egyptian years containing 29 February begin in an even Julian year (766 and 762). In the later part of the list, Egyptian years containing 29 February begin in an uneven Julian year (757 and 753). Such Egyptian years return to even years when the Egyptian New Year's Day reaches the beginning of the Julian year, which happens among others at the following juncture:

2 January 528	–	1 January 527	
2 January 527	–	1 January 526	
2 January 526	–	1 January 525	
2 January 525	–	*31 December 525*	[contains 29 February 525]
1 January 524	–	31 December 524	
1 January 523	–	31 December 523	
1 January 522	–	31 December 522	
1 January 521	–	*30 December 521*	[contains 29 February 521]
31 December 521	–	30 December 520	
31 December 520	–	30 December 519	
31 December 519	–	30 December 518	
31 December 518	–	*29 December 517*	[contains 29 February 517]

The year 521 BC is a leap year of 366 days. Two Egyptian 365-day years therefore begin in the same Julian year, the first on 1 January 521 BC and the second on 31 December 521 BC.

The assumption, again, is that the wandering motion outlined above proceeded without disruption from the early third millennium BC to the seventeenth century AD.

0.3. Two Problems with Writing a History of the 365-Day Year

The 365-day calendar was a focal point of the Calendars and Years session at the Notre Dame conference. A complete history of this calendar would be useful. This history spans almost all of human history, history being the period that begins when writing is first attested, from about 3000 BC onward. The Egyptian 365-day calendar is one of the longest-lived calendrical structures of all time. Egypt was the first nation to apply the number 365 in calendrical practice. And it was centuries ahead of everyone else. In fact, for nearly three thousand years, or three fifths of all of human history, they were all alone in applying the number 365 calendrically. A complete survey of the 365-day year would be a valuable contribution to chronology and its history. But there are two problems with presenting such a history here.

The first problem is that this history by itself would not definitively prove some basic assumptions regarding the Egyptian calendar, a central concern of the Notre Dame meeting. Among those basic assumptions is the complete regularity of its slow shift in relation to the seasons and in relation to the Julian calendar, its so-called wandering motion. Another basic assumption is the absence of positive evidence for rival calendars of similar length, say 365¼ days or 360 days, until after 30 BC. There is solid evidence for only one other Egyptian calendar, the lunar one, whose months last from about new moon to about new moon for an average length of 29.53 days, with lunar months varying in length between 29 days and 30 days, there being slightly more months of 30 days than months of 29 days, so as to achieve over time an average of 29.53 days.

The second problem is the large size of the topic, which spans many fields. A full treatment would require a study of monographic proportions. As was mentioned above, chronology is a story best told backwards. In pursuing the history of the 365-day year backward from early modern times, one encounters many milestones. I have identified at least twenty topics that would require a full treatment, as follows: (1) Ludwig Ideler's foundational book on ancient chronology and astronomy;[4] (2) the gradual replacement, for astronomical purposes, of the Egyptian 365-day year by the Julian year-count (from 4713 BC) in the seventeenth and eighteenth centuries AD; (3) the views of the chronologist Dionysius Petavius in the seventeenth century AD; (4) the views of the pioneering chronologist Joseph Justus Scaliger in the sixteenth century and early years of the seventeenth century; (5) the role of Paul Crusius' much underestimated *On the Epochs* of 1578; (6) Copernicus' understanding of ancient chronology, also in the sixteenth century; (7) the Byzantine and Arabic traditions of astronomy in the Middle Ages, including the chronologist Albiruni (around AD 1000); (8) work relevant to chronology accomplished at Alexandria in Egypt, from the third century BC to the fifth and sixth centuries AD, including Theon, Pappus, and of course Ptolemy's Almagest; (9) years in Demotic horoscopes and other assorted astronomical texts in hieroglyphic scripts; (10) the institution of the modified Egyptian year rival 365¼-day Alexandrian calendar in the late first century BC; (11) the chronology of the Ptolemies in all its complexity, from the third to the first centuries BC; (12) civil-lunar double dates in Aramaic papyri from fifth-century BC Egypt as well as other civil-lunar double dates dating to different periods in native Egyptian texts; (13) Assyrian king-lists of the late second and early first millennia BC; (14) the relevance of radiocarbon dating; (15) the Sothic dates of the second millennium BC; (16) Sothic-lunar dating, that is, the combining of Sothic dates and lunar dates for the purpose of chronology; (17) representations of the epagomenal days, that is, the five-day period at the end of the 365-day year; (18) the question of the role of the rising of Sirius and the Nile flood in the origin of the 365-day years; (19) early hieroglyphic annals such as the Palermo stone; (20) earliest attestations of civil months, such as those found in Djoser's step pyramid compound. Only some of these topics will be treated below.

0.4. Division of the 4500-Year History of the 365-Day Year into Periods

The principal division in the history of the Egyptian 365-day year is between the period before about 700–500 BC and the period after. It comes as no surprise that what is closer to us in time is better known. In the period from about 500 BC, I would myself consider the regular wandering course of the 365-day year to be certain beyond a doubt. The Egyptian calendar ceased being used by astronomers around AD 1700. From AD 1700 back to about 700–500 BC is a long time, more than 2000 years. Additional divisions will be useful. The following partition of this period into three intervals seems natural. The three intervals are demarcated by two signal developments.

The first development is the addition of a sixth epagomenal day at the very end of every fourth Egyptian civil year. This transformation of the Egyptian *civil* calendar is called the *Alexandrian* calendar. The Alexandrian year begins on Julian 29 August but on Julian August 30 after a leap year. Its average length is 365¼ days, that is, 3 × 365 + 366 divided by 4. The Alexandrian calendar was instituted some time between 30 BC and 22 BC. It remains uncertain, in spite of much discussion, when exactly the first extra day was inserted.

A couple of decades earlier, in 45 BC, another 365¼-day year was instituted by Julius Caesar in Rome. The Julian calendar is basically still the calendar we use today, except for a minor alteration decreed in 1582 by pope Gregory XIII, when the Julian calendar became the Julian-Gregorian calendar. In that year, 15 October immediately followed 4 October and ten days were omitted. As distinct from the Julian calendar, the Julian-Gregorian calendar omits three leap years in 400 years, those divisible by 100 but not by 400. Thus, AD 1700, AD 1800, and AD 1900 are not leap years. But AD 1600 and AD 2000 are. The Julian calendar hardly rivaled the Egyptian 365-day calendar inside Egypt. Soon after Rome annexed Egypt in 30 BC, Egypt acquired its own 365¼-day calendar, the afore-mentioned Alexandrian calendar. The Alexandrian calendar adds its intercalary day on Julian 29 August in the year before the one in which the Julian calendar has its intercalary day. Thus, AD 14/15 (29 August to 29 August), AD 19/20, and AD 23/24, and so on, are leap years in the Alexandrian calendar. AD 16, AD 20, and AD 24, and so on, are leap years in the Julian calendar.

The second signal development is the obsolescence and ultimate complete loss, in daily life, of the Egyptian 365-day civil calendar in favour of the Alexandrian calendar. This obsolescence is a process that took a couple of centuries. It would appear that the 365¼-day Alexandrian calendar became a serious rival of the Egyptian civil calendar soon after it was instituted some time in the period from 30 BC to 22 BC, especially in official contexts. The use of the Egyptian civil calendar in daily life seems to have declined strongly by the second century AD. Its use in daily life to any small degree cannot have extended beyond the fourth century AD. At least, no evidence has emerged to support such an extension.

The two signal developments that have just been described naturally divide the period from 700–500 BC to the early modern age into three intervals. In a first interval, lasting from about 700–500 BC to 30–22 BC, the 365-day year totally dominated. In the second interval, from 30–22 BC to the fourth century AD, hardly later, the Egyptian civil calendar was used alongside the Alexandrian calendar in daily life. In the third interval, from the fourth century AD at the latest until the modern age, the Egyptian civil calendar was used exclusively in astronomy and chronology, in which it retained a dominant role. The Egyptian civil calendar owed its enduring popularity in astronomy to its eminent simplicity. The cycle of 12 × 30 + 5 days makes counting intervals in days between two events easy. For example, the distance between a first event on Month 5 Day 13 of any given year and a second event on Month 10 Day 21 of some later year is at least exactly 5 months (10 – 5) plus 8 days (21 – 13). To 5 months 8 days, one adds the multiple of 365 days that corresponds to the number of *full* 365-day years that intervene between the two dates.

1. First Milestone in the History of the 365-Day Calendar (AD 1820s): Champollion Discovers the Seasonal Month Names in Hieroglyphic Sources

The Egyptian civil calendar was presumably always divided into twelve months of 30 days plus five added days exactly, for a fixed total of 365 days ($12 \times 30 + 5 = 365$). Over the course of Egyptian history, the twelve months are named in three ways, no more, no less.[5] The first two concern sets of twelve names. The two sets partly overlap. One set was derived from the other. I would call one set the theophoric set X and the other the theophoric civil set. There is variation in the names. Lists of both sets, with one variant for each name, are as follows:[6]

Theophoric Set X	Theophoric Civil Set
ṯḫ	*ḏḥwty*
mnḫt	*p n jpt*
ḥwt ḥr	*ḥwt ḥr*
kꜣ ḥr kꜣ	*kꜣ ḥr kꜣ*
šf bdt	*tꜣ ꜥbt*
rkḥ wr	*p n pꜣ mḫr*
rkḥ nḏs	*p n jmn ḥtp*
rnnwtt	*p n rnnwtt*
ḫnsw	*p n ḫnsw*
ḫnt ḫt	*p n jnt*
jpt ḥmt	*jpjp*
wpt rnpt or *rꜣ ḥr ꜣḫty*	*mswt rꜥ* or *rꜥ ḥr ꜣḫty*

In the third nomenclature, months are not identified by names as such but as first, second, third, or fourth month of one of three seasons of four months of 30 days or 120 days each. Importantly, this third way of naming months occurs *only in hieroglyphic sources*. It is not attested in any language of foreign origin spoken on ancient Egyptian soil, including Greek or Aramaic. Nor is it found in Coptic, the latest phase of Egyptian written mainly with the Greek alphabet. The hieroglyphic Egyptian names of the three seasons are *ꜣḫt*, *prt*, and *šmw*.

It was Jean François Champollion who discovered the naming of the months by the three seasons. He recounts the process of discovery in detail in an essay first read at the French academy in 1831 but published only in 1842.[7] The first steps in the discovery process date back to 1824 when Champollion was in Turin to study the collection of Egyptian monuments acquired by the king of Sardinia. Champollion wrote about his discovery in his Second letter to the Duc de Blacas. A comparison with a statue at the Louvre proved crucial. But soon after, Champollion stumbled upon the best confirmation of his discovery in lists of month names preserved on the walls of the Ramesseum temple in Thebes and of the Horus temple at Edfu. In those two lists, the seasonal names are listed next to the month-gods whose names are reflected in the two other sets of month names. Of these two other sets, the theophoric civil set is preserved in Greek and other languages and was therefore well-known to early modern scholars, including Champollion, even before the decipherment of hieroglyphic writing. Champollion already knew about the seasonal names before finding them at the Ramesseum and at Edfu, in contradiction with the following statement by Lepsius: "Champollion entdeckte zuerst die Monatsgötter in Theben und Edfu, und brachte sie mit den Monatsnamen in Verbindung in einer schönen Abhandlung, deren Inhalt, größtenteils wörtlich abgeschrieben, zuerst unter Salvolinis Namen publicirt wurde".[8]

Champollion presented his findings orally in a session of the Académie des inscriptions et belles-lettres on 18 March 1831. On 4 April 1831, he read a brief summary ("Note sommaire") of the same results at the Académie, immediately before Biot presented his own "Recherches sur l'année vague des Égyptiens". Champollion died about a year later on 4 March 1832. Both communications to the Academy remained unpublished for a decade and apparently could not be located. They resurfaced about a decade later and were published in 1842.[9] Meanwhile, in 1833, Salvolini had made Champollion's findings public without giving due credit (see Lepsius cited above).[10]

In his foundational *Aegyptens Stelle in der Weltgeschichte* (1845–1857), Bunsen appears quite impressed by Champollion's discovery of the division of the year into three seasons. Volume 4 of Bunsen's work concerns Egyptian chronology. It appeared in 1856, after Lepsius' pioneering book on the same subject of 1849.[11] Bunsen relies on the results of Lepsius' book. Volume 4, on chronology, begins with a plate depicting the following items: (1) Champollion; (2) the hieroglyphic names of the rulers of the Late period that led to Champollion's decipherment of the hieroglyphs; (3) the names of the seasonal months discovered by Champollion. Next is a poem of four (4!) pages extolling Champollion's discovery of the naming of months by seasons. Four lines are as follows:

> Nacht auch verhüllete bald das Geheimnis deutender Zeichen:
> "Viermal Wasser, und Sproß viermal, und Speicher auch viermal".
> Da enthüllete der Genius Dir, o unsterblicher Forscher,
> In der Monate Bild Kunde des Jahres und der Zeit.

Bunsen overrates the importance of Champollion's discovery of the seasonal month names, especially when he places this discovery on the same level as Champollion's decipherment of the hieroglyphic script and language.[12] The seasonal names are just one aspect of the Egyptian 365-day civil calendar. It is not as if Champollion discovered the existence of the Egyptian 365-day calendar itself, as Bunsen may seem to imply. The 365-day calendar was already well-known before Champollion. The knowledge of it never became obsolete because it was used in astronomy down to early modern times. When Ideler described this calendar in his manual of chronology of 1825–1826, the Egyptian seasonal month names were still unknown to him. Champollion had just discovered them in 1824.[13]

2. Second Milestone in the History of the 365-Day Calendar (AD 1806): Ideler First Fully Evaluates the Chronological Data in Ptolemy (Second Century AD), Thus Establishing the Foundation of Modern BC Chronology

Once the Egyptian wandering year was abandoned in daily life, at the latest probably by the fourth century AD, it lived on in astronomy. The 365-day calendar was known and understood down to modern times without interruption because Ptolemy used it in his Almagest (second century AD), antiquity's greatest work of astronomy. There is a continuous line of transmission and study of the Almagest from when it was first written down to the present time.

Ptolemy counted Egyptian 365-day years in continuity according to the Era from Nabonassar or according to the Era of Philip. An era is a way of counting years in perpetuity from a Year 1. Counting years by eras is distinct from the normal manner of counting of years in antiquity, namely by reign, with the count starting over with each new ruler. Day 1 of an era is called its epoch. In its non-technical meaning, "epoch" otherwise rather refers to a large interval of time. The epoch of the Era of Nabonassar is 26 February 747 BC. Year 425 of the Era of Nabonassar is the same as Year 1 of the Era of Philip. The epoch of the Era of Philip is therefore 12 November 324 BC. Alexander the Great died in the Egyptian 365-day year lasting from 12 November 324 BC to 11 November 323 BC.[14] The Era of Philip is counted from the year of Alexander's death as Year 1.

Ptolemy counted Egyptian years not only by the Era of Nabonassar but also by kings' reigns each lasting an integer number of 365-day years. This succession of reigns is known as Ptolemy's Royal Canon.[15] The Canon is nothing but a subdivision of the Era of Nabonassar into reigns. But the Era and the Canon have sharply different histories of transmission. The Era remained known and understood from antiquity to the present. Knowledge of the Canon was lost in the Latin West. Correct versions of the Canon first emerged in the West in Greek manuscripts in the early seventeenth century AD.

Christiaan Huygens (1629–1695) still used the Egyptian civil year for dating astronomical events at the end of the seventeenth century.[16] Egyptian years and the Era of Nabonassar are now no longer used in astronomy. What replaced them is the Julian era, a count of years proposed by Joseph Scaliger in the first edition of his *De emendatione temporum* (1583). Scaliger called the year Julian because its year is the Julian year of 365¼ days instituted by Julius Caesar in 45 BC. Indeed, in his *De emendatione*, he writes, "Iulianam vocavimus, quia ad annum Iulianum duntaxat accommodata est".

Year 1 of Scaliger's Julian era is 4713 BC. Scaliger obviously did not refer to it as 4713 BC. Counting years BC was introduced in the seventeenth century only. 4713 is not a random number. It is the result of making three existing historical cycles—one of 15 years, one of 19 years, and one of 28 years—begin at the same time well before earliest known history. The problem is one of indeterminate algebra. 4713 is its necessary solution.[17]

There is a widespread, if not universal, misconception found in all the handbooks that Scaliger also counted *days* from 1 January 4713 BC. This is not the case. Counting days from 1 January 4713 BC is a product of the nineteenth century. The days are counted as j.d. ("Julian day"). J.d. numbers are *cardinal* numbers, not *ordinal* numbers. For example, 1 January 4713 BC is the day that lasts 24 hours from j.d. 0.00 (midnight between 31 December 4714 BC and 1 January 4713 BC) to j.d. 1.00 (midnight between 1 January 4713 BC and 2 January 4713 BC). J.d. 0.50 is noon or 12:00PM of 1 January 4713 BC. Scaliger created his "Julian" *year* count from 4713 BC for the benefit of *historians*. It was *astronomers* who, a couple of centuries later, derived the j.d. *day* count from the year count of the Julian era. I am disregarding the fact that astronomers have often counted days from noon to noon, because that allows them to combine the collective astronomical events of a single night as part of the same date.

This shift from dating astronomical events by the Egyptian 365-day calendar to dating them by "j.d." numbers parallels a deeper shift in the evolution of astronomy. Until Newton (1642–1727), Ptolemy's Almagest was the foundation of astronomy. The Almagest dates astronomical events by Egyptian years. From Newton onward, the nature of astronomy fundamentally changed. Neugebauer has described this change as follows.

> For the history of mathematics and astronomy the traditional division of political history into Antiquity and Middle Ages is of no significance. In mathematical astronomy ancient methods prevailed until Newton and his contemporaries opened

> a fundamentally new age by the introduction of dynamics into the discussion of astronomical phenomena. One can perfectly well understand [Newton's] "Principia" without much knowledge of earlier astronomy but one cannot read a single chapter in Copernicus or Kepler without a thorough knowledge of Ptolemy's "Almagest". Up to Newton all astronomy consists in modifications, however ingenious, of Hellenistic astronomy.[18]

The focus is presently on Ptolemy's Almagest. Neither BC chronology in general nor the role of the Egyptian 365-day year in the study of ancient history in specific could have been adequately understood until the chronological data in the Almagest had been analysed. Ludwig Ideler (1766–1846) produced such an analysis in his *Historische Untersuchungen* of 1806. It is little realized that this book is the foundational work of modern BC chronology. In 1825–1826, Ideler published his monumental handbook of chronology. But already two decades earlier, he had fully articulated where BC chronology begins. An adequate understanding of ancient chronology begins with knowing all that there is to know about how chronological data are manipulated in Ptolemy's works on astronomy. In his *Untersuchungen*, Ideler first attained such knowledge. He collected all the chronological material found in Ptolemy and once and for all established its inner coherence and its veracity.

Joseph Justus Scaliger (1540–1609) is often portrayed as the pioneer of the modern study of chronology. Ideler himself states about Scaliger's Julian era (from 4713 BC) that "order and light came into chronology only with [its] introduction".[19] However, if Scaliger's *De emendatione* is the foundation of *all* of the modern study of chronology,[20] then Ideler's *Untersuchungen* is no doubt the foundation of the modern study of *BC* chronology.

One key ingredient was still missing from Ideler's BC chronology: direct access to many BC and early AD sources. The following four developments in the nineteenth and early twentieth centuries AD enabled such access, but only after Ideler had completed his work on chronology: (1) the decipherments of the hieroglyphic and cuneiform scripts; (2) the rise of papyrology; (3) the decipherment of Babylonian astronomy; and (4) the discovery of Aramaic papyri in Egypt containing double dates. Owing to such developments, a proper appreciation of the use of the Egyptian 365-day year in daily life prior to the time when it was used by astronomers became possible.

3. THIRD MILESTONE in the History of the 365-Day Calendar (AD 1578): Crusius Puts Ptolemy's Era of Nabonassar in Order

Ideler's *Untersuchungen* of 1806 share much with Crusius' *Liber de epochis* "Book on the Epochs" of 1578.[21] In regard to external appearance, both are slim books. Yet, both are milestones in the modern study of chronology. Both authors were trained astronomers. Both placed a proper understanding of the chronology of Ptolemy's Almagest with its Egyptian 365-day year and its Era of Nabonassar at the very beginning and origin of ancient chronology and therefore also of ancient history. Both works were soon overshadowed by large volumes that appeared soon after, in the case of Crusius by Scaliger's *De emendatione temporum*[22] and in the case of Ideler by Ideler's own *Handbuch*.[23] There is not much awareness in the field of ancient history at large that AD 1578 and AD 1806 are so paramount as milestone dates in the study of antiquity.

Receding into the past from Ideler's *Untersuchungen* (1806), one first encounters the activity of chronologers in the eighteenth and seventeenth centuries AD. Much more analysis is needed of the evolution of chronology in those decades. The task at hand will involve the perusal of many heavy tomes typically written in Latin and found mainly in rare book libraries. There is so much more to be known about who knew what or understood what or had access to which sources at which time in this period. It is the period in which the Egyptian 365-day year ceased being used actively in astronomy. It is also the period in which the count of "years BC" came into vogue, presumably first in Petavius' *Opus de doctrina temporum* (1627–1630).[24] Also in that period, probably in AD 1740 to be precise, Jacques Cassini first used the year count including a year 0.[25] In that count, Year 0 is 1 BC, Year –1 is 2 BC, Year –2 is 3 BC, and so on.

What matters presently is that the Egyptian 365-day year hardly played a role in the study of ancient history in Crusius' time. To date Greece and Rome, chronologers relied on tools such as Olympiads, lists of archons, lists of consuls, and the year count A.U.C. (*ab urbe condita* "from the foundation of Rome", that is, from 753 BC). Not enough was known about the ancient Near Eastern civilizations of Mesopotamia and Egypt from direct sources for the Egyptian 365-day calendar to be of great use as a dating tool for historians.

The limited role of the Egyptian 365-day year can also be observed in the chronological work of two pioneers of the modern study of chronology, Joseph Justus Scaliger (1540–1609) and Dionysius Petavius (1583–1652). Scaliger is the true originator of modern chronology. His erudition was massive. But he was occasionally prone to fantasy. Thus, he even thought he could disprove the precession of the equinoxes.[26] In his monumental *Opus de doctrina temporum* (1627–1630), whose Foreword is addressed to Cardinal de Richelieu, Petavius follows Scaliger step by step, adopting and castigating Scaliger's results to such equal degrees that Rühl wondered "whether [Petavius] admired or detested Scaliger more".[27]

The present focus is on the 365-day year and the Era of Nabonassar. Backward from Scaliger, Ptolemy's Canon was not a factor. Correct versions of the Canon emerged only after Scaliger's death. In properly evaluating the Era of Nabonassar, Scaliger owed much to a little book by Paul Crusius. "[I]t was an epochal book", writes Grafton, "in quality as well as subject; without it the *De emendatione* would never have taken its final form".[28] Grafton has done much to rehabilitate Crusius' slender volume "on the epochs" (that is, on the starting dates of eras) as a veritable milestone in the evolution of chronology.[29] Paulus Crusius was professor of astronomy and mathematics in Jena. His book on the epochs was published posthumously in 1578.

The study of Manilius' work on astronomy had whetted Scaliger's appetite for matters chronological in the late 1570s. But in 1580–1581, he became hopelessly stuck. In May of 1581, two years before the publication of his foundational *De emendatione*, we find him eagerly trying to obtain a copy of a book "on correcting epochs in chronology" "published in Basel by a German".[30] He twice writes to his friend Claude Dupuy requesting a copy of the book and finally receives one in the summer of 1581. "[H]is encounter with Crusius' able little book proved decisive," writes Grafton. "[Crusius] raised the stakes of—and the price of entrance into—chronological debate. And in this case, as often before, the challenge of a brilliant predecessor forced Scaliger to reconceive his own work. In the end, he built his own structure on the foundations Crusius had laid, thus making them unrecognizable and consigning them to oblivion".[31]

Crusius' accomplishment does not pertain to the structure of the Egyptian 365-day year. How difficult can it be to understand a perpetual cycle of 12 × 30 + 5 days? The only possible complication regarding this simple structure is the relation between the 365-day wandering civil calendar and another Egyptian calendar, the 365¼-day Alexandrian calendar, which is derived from the wandering calendar by the insertion of an intercalary day every four years. The difference between the two calendars is in fact a topic of discussion in an imaginary dialogue between a teacher and a confused student found in a manual of chronology published two years before Crusius' own book, namely Erasmus Oswald Schreckenfuchs' *Opus posthumum* (Basel, 1576), at sig. [HHH 4].[32] The student had read that the Egyptian New Year's Day is fixed at *quarto Calendas Septembris* "29 August".[33] But he apparently cannot reconcile this report with his assumption that the Egyptian New Year's Day is *not* fixed. Julian 29 August (30 August after an Alexandrian leap year with an intercalary Julian 29 August) is indeed the New Year's Day of *an* Egyptian calendar, but not of the wandering calendar but rather of the so-called Alexandrian calendar. The intercalary day is all that differentiates the wandering calendar from the Alexandrian calendar. The Alexandrian calendar, instituted in the 20s BC, and the Julian calendar, instituted by Caesar in 45 BC, both have leap years of 366 days. Their average length is therefore 365¼ days (365 + 365+ 365 + 366 divided by 4). The Egyptian civil calendar does not have an intercalary day and therefore wanders in relation to both the Alexandrian and the Julian calendars at a rate of one day in four years. In Schreckenfuchs' imaginary dialogue, the teacher clarifies the difference between the wandering Egyptian civil calendar and the fixed Alexandrian calendar, with reference to Ptolemy's Almagest. The student had read part of the Almagest but only poorly understood it. In addition, he had read the epitome, or summary, of the Almagest by Johannes Müller (1436–1476) of Königsberg. Müller is better known by the Latinization of his provenance, Königsberg (not the present Kaliningrad, Kant's Königsberg, but rather a small city in Bavaria) as Regiomontanus ("he of King's Mountain").

Crusius' accomplishment pertaining to the proper understanding of the Era of Nabonassar was all the more significant as the Era had come to be misunderstood owing to an error by Copernicus, whose understanding of ancient chronology is otherwise an important episode in its own right in the history of the Egyptian 365-day year. The error came about as follows.

Two ways of counting years in perpetuity from a Year 1, that is, counting years as an era, are crucial to understanding the history of the first millennium BC. The first era is the Olympiads, four-year cycles of which each year begins and ends with or sometime around the Olympic games in the summer. The second era is the Era of Nabonassar. There is an important difference between the two eras. The Olympiads were actively used for dating events already in antiquity, though hardly as early as the eighth century BC. The Era of Nabonassar was always an artificial count. It was designed for use by astronomers only. However, the Era of Nabonassar has proved to be of fundamental importance to historians as well. The epoch of the Olympiads is some time in the summer of 776 BC. Olympiad I.1, or the first year (1) of the first Olympiad (I), is therefore the first year of the Olympiadic Era; Olympiad II.1 is its fifth year, and so on.

Copernicus (1473–1543) was familiar with both eras. Swerdlow and Neugebauer have described Copernicus' treatment of ancient chronology.[34] Copernicus was obviously well acquainted with the Egyptian 365-day calendar. In spite of its geocentric bias, the Almagest remained the foundation of astronomy as late as in Copernicus' time. Copernicus' own copy of Ptolemy's Almagest has been preserved, complete with his handwritten annotations. Copernicus correctly dates the epoch of the Era of Nabonassar to 26 February 747 BC. However, he erroneously places the beginning of the Olympiadic Era a year too late, in the summer of 775 BC, choosing 1 July of that year as an artificial beginning. Swerdlow and Neugebauer have sought to explain this one-year deviation by assuming that Copernicus relied on Eusebius' chronology.[35] Whatever may be of the matter, from 1 July 775 BC to 26

February 747 BC, there are exactly 10,102 days, as Copernicus too counts. That is 27 full Egyptian years of 365 days plus 247 days.

The present concern is with the Egyptian 365-day year. Copernicus' error regarding the beginning of the Olympiadic Era—placing it in the summer of 775 BC instead of in the summer of 776 BC—spawned a second error regarding the beginning of the Era of Nabonassar in the writings of others who adopted Copernicus' 775 BC. The year of the Era of Nabonassar is the Egyptian 365-day year. As a result of this second error, confidence in the Egyptian 365-day year as a tool for measuring time was shaken. The structure of the 365-day calendar seemed eminently simple. Yet dates of astronomical events by the Era of Nabonassar could not be verified if one assumed that the Egyptian year had shifted with consistency by one day in four years in relation to the Julian year.

The second error came about as follows. As was noted above, Copernicus assumed that 27 years plus 247 days, that is, *less than* 28 full years, separated the beginning of the first Olympiad from the beginning of the Era of Nabonassar. This assumption was in contradiction with the general belief at the time that the interval between the two epochs was *more than* 28 full years, as is in fact the case. This interval of 28-plus years had been obtained already before Copernicus by the following reasoning.[36]

Two observations apply. The first observation is that, by the sixteenth century AD, it was generally inferred from the ancient sources that Alexander had died in the beginning of the 114th Olympiad, a four-year period that lasts from the summer of 324 BC to the summer of 320 BC. In actual fact, Alexander died on 11 June 323 BC,[37] that is, near the end of the first year of the 114th Olympiad. Olympiad CXIV.1 lasts from the summer of 324 BC to the summer of 323 BC. 113 Olympiads of four years, or 452 years (113 × 4), therefore precede the summer of 324 BC back to the beginning of the Olympiadic Era, that is, the summer of 776 BC (324 + 452).

The second observation is that Year 1 of the Era of Philip, also called the Era from the Death of Alexander, is Year 425 of the Era of Nabonassar. 424 Egyptian 365-day years therefore separate the beginning of the Era of Nabonassar from the beginning of the Era of Philip. 424 Egyptian years are about 100 days shorter than 106 Olympiads.

These two observations involve two intervals. The first interval of 452 years is the distance between the beginning of the Olympiadic Era and the beginning of the Era of Philip. The second interval of 424 years is the distance between the beginning of the Era of Nabonassar and the beginning of the Era of Philip. The result of subtracting 424 from 452 is 28. The number 28 somehow represents the interval between the beginning of the Olympiadic Era and the beginning of the Era of Nabonassar. In reality, we know now that the interval between one beginning (summer of 776 BC) and the other (26 February 747 BC) is roughly as long as, if not longer than, 28.5 solar years. But how to interpret the number 28?

Two factors effect that the interval between the beginnings of the two eras is *more than* 28 solar years. The first factor is that the years of the 452-year interval and the years of the 424-year interval differ in length. The 452 years are solar years—strictly speaking Olympiadic years that consist of either 12 or 13 lunar months and over time average out to the length of the solar year of 365.2422 years. The 424 years are shorter, Egyptian years of exactly 365 days.

If one counts back 452 solar years and 424 solar years from a single point in time, one will obtain two points in time that are separated by an interval of exactly 28 solar years. However, in the present case, the years of the 424-year interval are shorter and will therefore extend less far into the past. Accordingly, the difference between the two points in time will increase to *more than* 28 solar years (452 – 424). Each of the 424 years is on average a little under a quarter day, or about 0.2422 days, shorter than a solar year. The interval between the beginnings of the two eras will therefore increase by a little more than 100 days (424 × about 0.2422) above 28 solar years.

The second factor is that the 452-year interval and the 424-year interval are counted backward from different points in time. The 452 years are counted back from the summer of 324 BC. Olympiadic years begin in the summer. The 424 years are counted back from a few months *later*, 12 November 324 BC, or Day 1 of Year 425 of the Era of Nabonassar or Day 1 of Year 1 of the Era of Philip. This factor increases the interval between the beginnings of the two eras by an additional three to four months above 28 solar years, corresponding to the interval between the summer of 324 BC and 12 November 324 BC.

In sum, the two factors combined increase the distance in time between the beginnings of the two eras by half a year or more above 28 solar years. This is in fact the correct interval as we know it.

Copernicus erred in placing the beginning of the Olympiadic Era in the summer of 775 BC, a year too late (see above). It was generally—and, as it happens, correctly—assumed at the time that the interval between the beginning of the Era of Nabonassar and the Olympiadic Era was at least 28.5 years (see above). Anyone adopting Copernicus' error would count at least 28 solar years and six months from July/August of 775 BC to obtain the beginning of the Era of Nabonassar and therefore place the beginning of the Era of Nabonassar in early 746 BC. It is then a small step to place the beginning of the Era of Nabonassar exactly 365 days, or one full Egyptian year, later than 26 February 747 BC, its real beginning. 26 February 746 BC is where the eminent Flemish geographer Gerhard Mercator (1512–1594) places the beginning of the Era of Nabonassar in his *Chronologia* (Cologne, 1569).[38] It should be added that the convention of extending the Julian calendar back to before AD did not yet exist. Mercator did therefore not refer to the day in question as 26 February 746 BC.

Ptolemy uses the Era of Nabonassar to date astronomical events such as eclipses. In dating astronomical events, a deviation of just one hour can already produce a defective date for an event. Mercator placed the beginning of the Era of Nabonassar a full Egyptian year too late. It did not take him long to notice that Ptolemy's dates for astronomical events could not be verified. Mercator went as far as postulating a type of lunar motion that had not yet been discovered in order to account for the inconsistencies in Ptolemy's dates.

It was Crusius who, once and for all, firmly placed the beginning of the Olympiads in the summer of 776 BC instead of Copernicus' 775 BC and the beginning of the Era of Nabonassar on 26 February 747 BC instead of Mercator's 746 BC. Crusius also first explicitly gives Ptolemy full credit for making a precise chronology of antiquity at all possible.[39] His main objective was to restore the beginning of the Olympiadic Era to 776 BC. Crusius systematically gathered and analysed dates of eclipses according to Olympiads found in classical sources. There could no longer be any doubt: Olympiad I.1 began in the summer of 776 BC.

A few years later, Scaliger published his own monumental work on chronology, drowning out Crusius' accomplishments. What earned Scaliger his reputation as pioneer of the modern study of chronology is his unparalleled codicological, linguistic, and philological skills. These skills allowed him to digest and combine vast amounts of esoteric source material in multiple languages into one coherent whole. But Scaliger's skills in astronomy did not equal Crusius' or Petavius'.

4. FOURTH MILESTONE in the History of the 365-Day Calendar (Fifth Century BC): Babylonian and Egyptian Dates as a Rule Match in Double Dates from Aramaic Papyri Found in Egypt

4.1. The Egyptian 365-Day Year Backward from Copernicus and Crusius

The preceding sections 1, 2, and 3 concern the insightful efforts on the part of three scholars, Champollion, Ideler, and Crusius, that proved fundamental to our understanding of the Egyptian 365-day calendar and to its role in the study of ancient history. None of their contributions are voluminous in extent. As a result, these contributions may not appear as prominent as they deserve to be. Crusius firmly fixed the course of the Era of Nabonassar's Egyptian 365-day year by first fixing the beginning of the Olympiadic Era through a comprehensive analysis of eclipse reports in classical sources. Ideler put Ptolemy's Almagest, which dates astronomical events according to the Egyptian 365-day year, firmly at the origin of the logical structure of ancient chronology. Champollion first revealed the 365-day calendar in hieroglyphic sources; months were counted as first, second, third, or fourth month of one of three seasons of four months.

As one moves backward from the times of Champollion, Ideler, and Crusius, the question arises as to how we can know whether the course of the 365-day calendar is indeed where one expects it to be at all times. Back to about 400 AD, the calendar was used only artificially for astronomical purposes. The uninterrupted chain of transmission and proper understanding of Ptolemy's Almagest from when it was written down in the second century AD until Copernicus and Crusius in the sixteenth century AD guarantees that the wandering course of the Egyptian year was regular. During most of this period of over a millennium, the dominant traditions of astronomy were Byzantine and Arabic.[40] Both traditions included a full understanding of Ptolemy's opus and therefore of the mechanics of the 365-day wandering year.

Backward from about AD 400, one enters into the period when the 365-day calendar was also used in daily life in Egypt. What guarantee is there that the Egyptian calendar of daily life and the Egyptian calendar of the astronomers at all times ran parallel? The best evidence back to about 500 BC is a set of double dates in Aramaic papyri from Egypt of the fifth century BC. In these double dates, a Babylonian date is equated with an Egyptian date. The double dates involve the use of the calendar in daily life. If the Egyptian calendar was where we expect it to be in both the fifth century BC and the fourth century AD, we can safely assume that it was where we expect it to be everywhere in between. It was known throughout those centuries at the time why the Egyptian year wandered and what could be done about it. Nevertheless, the year was allowed to wander with absolute regularity for a millennium. If the 365-day year was completely regular with certitude from about 500 BC to about AD 400, then why not also before 500 BC? Still, in section 5 below, this assumption will not be taken for granted. Rather, it will be shown that the independently obtained chronologies of Egypt and West Asia back to about 1500 BC confirm the assumption.

4.2. Completely Preserved Aramaic Double Dates from Fifth-Century BC Egypt

A double date is an equation between two dates given in a document, as a rule at its beginning. In Aramaic papyri from Egypt dating to the fifth century B.C., a number of double dates have been preserved. As immigrants from West Asia, the scribes of these papyri used a calendar modeled after, if not nearly identical to, the Babylonian calendar. As inhabitants of Egypt, they also knew how to date by the local Egyptian 365-day civil calendar. In many cases, they dated a document both by their own Babylonian calendar and by the calendar of their host-country. The two dates equated with one another in a double date are therefore a Babylonian date and an Egyptian date. The two calendars obviously differed much in structure. The Babylonian calendar consists of years of either 12 or 13 lunar months of either 29 or 30 days. In the Egyptian calendar, each year is exactly 365 days long and consists of 12 months of 30 days plus five added days.

Among the Aramaic double dates, fourteen are completely preserved and seem free of any philological problems that might interfere with attempts at chronological interpretation. The fourteen equations are shown in table 1.[41] This table contains fourteen double dates that produce equations between two Julian dates in columns (4) and (5). The two Julian dates have been obtained *independently* from one another. The first Julian date in column (4) is the Julian equivalent of the Egyptian component of the double date, on the assumption that the wandering motion of the Egyptian civil calendar was absolutely consistent. The second Julian date in column (5) is the Julian date of the closest conjunction, that is, the time when the moon is exactly between the sun and the earth and can therefore not be seen from the earth.[42] If one compares columns (4) and (5), it appears that the two dates of each pair of dates fall close to one another. In other words, there is a match between the two Julian dates. The essential data, extracted from the table, are as follows.

	Julian Date 1 (daylight of Day 1 of the lunar month)	Julian Date 2 (new moon or conjunction)
No. 1	**4 Mar** 471	**1 Mar** 471 (2:41PM)
No. 2	**26 Aug** 471	**24 Aug** 471 (6:13PM)
No. 3	**12 Dec** 457	**16 Dec** 457 (8:06AM)
No. 4	**17 Jun** 451	**16 Jun** 457 (1:59PM)
No. 5	**16 Nov** 446	**16 Nov** 446 (5:17AM)
No. 6	**13 Aug** 440	**12 Aug** 440 (7:16PM)
No. 7	**8 Sep** 437	**7 Sep** 437 (12:02PM)
No. 8	**6 Oct** 434	**4 Oct** 434 (8:43AM)
No. 9	**24 May** 427	**22 May** 427 (4:58AM)
No. 10	**17 Sep** 416	**15 Sep** 416 (5:32AM)
No. 11	**14 Dec** 416	**12 Dec** 416 (11:52PM)
No. 12	**18 Jan** 410	**17 Jan** 410 (2:54AM)
No. 13	**2 Nov** 404	**1 Nov** 404 (10:41AM)
No. 14	**18 Feb** 402	**16 Feb** 402 (8:14PM)

The matches between the two Julian dates may be defined more precisely as follows. I am disregarding for the moment the problem as to when exactly in relation to new moon or conjunction the lunar months in question began. More on this follows below. What matters at present is the rough observation that the beginnings of the months consistently *fell close* to new moon. New moon is a computed item and therefore suitable for a consistent and objective comparison. One new moon (in no. 1) is three days earlier than Day 1 of the month. Six new moons (in nos. 2, 8, 9, 10, 11, and 14) are two days earlier than Day 1. Five new moons

	Text	(1) Egyptian Component (Civil)	(2) Babylonian Component (Lunar)	(3) (1) = (2) converted for Babylonian Day 1	(4) Daylight of Egyptian date in (3)	(5) New moon close to (4)
1	C3.8	9 Choiak Year 15	24 Adar Year 14 of Xerxes I	16 Hathyr = 1 Adar	4 Mar 471	1 Mar 471 (2:41PM)
2	B2.1	28 Pachons Year 15 of Xerxes I	18 Elul	11 Pachons = 1 Elul	26 Aug 471	24 Aug 471 (6:13PM)
3	B3.1	4 Thoth Year 9 of Artaxerxes I	7 Kislev	3 epagomenal = 1 Kislev	12 Dec 457	16 Dec 457 (8:06AM)
4	B3.2	25 Phamenoth Year 14 of Artaxerxes I	20 Sivan	6 Phamenoth = 1 Sivan	17 Jun 451	16 Jun 451 (1:59PM)
5	B2.7	10 Mesore Year 19 of Artaxerxes I	2 Kislev	9 Mesore = 1 Kislev	16 Nov 446	16 Nov 446 (5:17AM)
6	B2.8	19 Pachons Year 25 of Artaxerxes I	14 Ab	6 Pachons = 1 Ab	13 Aug 440	12 Aug 440 (7:16PM)
7	B3.4	9 Payni Year 28 of Artaxerxes I	7 Elul	3 Payni = 1 Elul	8 Sep 437	7 Sep 437 (12:02 PM)
8	B3.5	25 Epeiph Year 31 of Artaxerxes I	25 Tishri	1 Epeiph = 1 Tishri	6 Oct 434	4 Oct 434 (8:43AM)
9	B3.6	7 Phamenoth Year 38 of Artaxerxes I	20 Sivan	18 Mecheir = 1 Sivan	24 May 427	22 May 427 (4:58AM)
10	B3.9	22 Payni Year 8 of Darius II	6 Tishri	17 Payni = 1 Tishri	17 Sep 416	15 Sep 416 (5:32AM)
11	B2.10	12 Thoth Year 8	3 Kislev Year 9 of Darius II	10 Thoth = 1 Kislev	14 Dec 416	12 Dec 416 (11:52PM)
12	B2.11	9 Hathor Year 13	24 Shebat Year 14 of Darius II	16 Phaophi = 1 Shebat	18 Jan 410	17 Jan 410 (2:54AM)
13	B3.10	29 Mesore Year 1 of Artaxerxes II	24 Marcheshvan	6 Mesore = 1 Marcheshvan	2 Nov 404	1 Nov 404 (10:41AM)
14	B3.11	8 Choiak Year 3 of Artaxerxes II	20 Adar	19 Hathyr = 1 Adar	18 Feb 402	16 Feb 402 (8:14PM)

TABLE 1. Completely Preserved Double Dates in Aramaic Papyri from Egypt

(in nos. 4, 6, 7, 12, and 13) are one day earlier. One new moon (in no. 5) is on the same day. And one new moon is four days later (no. 3). In sum, in 12 out of 14 dates, the Julian date of conjunction is either the same day or one or two days earlier. Considering these matches as pure coincidence would be very difficult. The probability that they are a coincidence can be formulated mathematically. Such a mathematical interpretation follows below.

But why the variation from zero to two days? Such variation is in fact altogether expected with lunar dates. Lunar dates are dependent on observation of the moon. Observation of the moon itself involves many factors, too many and too complex to summarize here. The factors include the weather and the location of the observer. At all events, the variation does not in the least surprise.

Ancient months were mostly no doubt lunar. But as to when these lunar months exactly began, it is still widely assumed nowadays that they began with first crescent sighting. The first crescent is first visible above the western horizon soon after sunset one to two days after conjunction. If the first crescent was the marker of the beginning of the month, then *daylight of Day 1 of the lunar month ought to be the daylight period that immediately follows the evening of first crescent visibility.* I have elsewhere objected to this seemingly universally accepted view.[43] To the limited extent that their beginnings can be determined, ancient lunar months began a little too early to allow the first crescent to be seen. Instead of first crescent visibility, I proposed an "untidy but natural alternative". Much support for first crescent sighting derives from the absence of a tidy alternative. But is there need for such an alternative? Most peoples of all places and all times, from Micronesia to Lapland and from Siberia to the Andes use or have used lunar calendars. None of these calendars operate by rigorous rules. Why would ancient Near Eastern lunar calendars be different? The average astronomical lunar month is about 29.53059 days long. By using months of 29 days and 30 days, one can easily keep up with the lunar phases in the sky. If the first crescent is sighted too early, before 29 days have passed, or too late, a couple days after lunar Day 1, the calendrical authorities that decide for village, town, or country can easily correct the problem by their specific choices of 29-day and 30-day months. An untidy system does not need to be ineffective as the months pass by. Looking out for the first crescent still fulfills a function. It serves as a check that the astronomical month and the calendrical month have not drifted too far apart. By this theory, precise calculations of lunar visibility are mostly superfluous in the modern interpretation of ancient lunar calendars except for lunar calendars used by astronomers.[44]

4.3. Comments on Double Dates No. 3 and No. 10

No. 3 and No. 10 each pose a problem of their own. No. 3 clearly stands out among the 14 double dates listed above. It is the only date in which the lunar month begins before the day of conjunction, and four days before conjunction at that. As regards No. 10, lunar Tishri is the lunar month that precedes the lunar month Tishri of the well-established 19-year lunar cycle used by astronomers at Babylon. Accordingly, Tishri would have fallen a month earlier at Assuan, in the deep south of Egypt, than it did at Babylon. Can this be? Evidence in defence of such a possibility will be derived below from the Aramaic papyrus Egyptian Museum Cairo no. 3652 = JdÉ 37114.

4.3.1. Double Date No. 3

Double Date No. 3 at face value yields 12 December 457 BC as lunar Day 1 and 16 December 457 BC as day of the closest conjunction (by Porten's emendation, 7 December 456 BC is daylight of lunar Day 1 and 5 December 456 BC is day of the closest conjunction). 4 Thoth is 18 December in both 457 BC and 456 BC. Since 4 Thoth is equated with lunar 7 Kislev, lunar 1 Kislev is 12 December in both 457 BC and 456 BC. The conjunction closest to 12 December 457 BC is on 16 December at about 8:06AM. The conjunction closest to 12 December 456 BC is on 5 December 457 BC at about 11:06PM. Neither conjunction is sufficiently close to lunar Day 1 for a satisfactory match.

But because the conjunction in 457 BC is closer, 457 BC is given in table 1 in section 4.2. Even so, the lunar month of the double date then begins four days before the day of conjunction. This four-day gap deviates from what is the norm. The text is well preserved. One might declare the double date unusable and reduce the number of double dates from 14 to 13. Errors do occur in ancient dates. But is there an error in no. 3 and is there a plausible explanation for a possible error?

Porten has plausibly suggested both an error and an explanation for the error.[45] The error is that 4 Thoth is mistakenly written for fourth epagomenal day. The explanation for the error is that the scribe failed to count the five epagomenal days at the end of the 365-day year. In both 457 BC and 456 BC, the fourth epagomenal day falls on 13 December. Which of the two years is more probable or suitable, 457 BC or 456 BC? A comparison with lunar conjunctions provides the answer.

If the Egyptian date of no. 3 is indeed the fourth epagomenal day and hence falls on 13 December in either 457 BC or 456 BC, then lunar 7 Kislev also falls on 13 December in either 457 BC or 456 BC. The Egyptian date and the Babylonian date are after all equated with one another in the double date. If 7 Kislev is 13 December in either 457 BC or 456 BC, then 1 Kislev is 7 December (13 – 6 = 7) in either 457 BC or 456 BC. 1 Kislev is Day 1 of a lunar month. One reasonably expects lunar Day 1 to fall close to conjunction. Is this the case in either 457 BC or 456 BC?

The conjunction closest to 7 December in 457 BC is on 16 December 457 BC at about 8:06AM. That is about 9 days later than 7 December. The next closest conjunction is on 16 November 457 BC at about 9:05PM. That is about 21 days earlier. The year 457 BC is clearly not suitable to Porten's emendation.

The conjunction closest to 7 December in 456 BC is on 5 December 456 BC at about 11:06PM. The relation between 7 December 456 BC as the Julian date of Day 1 of the lunar month Kislev and 5 December 456 BC as the Julian date of conjunction is well within the norm exhibited by the other 13 completely preserved double dates listed in 4.2. The year 456 BC is therefore eminently suitable if one accepts Porten's emendation. In a sense, this happy outcome is an argument in favour of Porten's emendation.

The year date of no. 3 is Year 9 of Artaxerxes I. Babylonian and Egyptian regnal years begin at different times. Years of the same number therefore only partly overlap. For part of the year, the Babylonian year number is one *higher* than the Egyptian year number. We know that Artaxerxes I's Egyptian regnal Year 9 lasted from 15 December 457 BC to 14 December 456 BC. And we know that his Babylonian Year 9 lasted from April 456 BC to April 455 BC. According to Porten's emendation, the date of the document in which double date no. 3 appears is 13 December 456 BC. This date suitably falls both inside the Egyptian regnal Year 9 and inside the Babylonian regnal Year 9. Indeed, the date falls in that portion of the solar year in which the Egyptian year and the Babylonian year have the same number. The year date therefore does not contradict Porten's emendation. The Egyptian regnal Year 10 begins two days later on 15 December 456 BC. The Babylonian regnal Year 10 begins a couple of months later in the spring.

Without emendation, the Egyptian date of double date no. 3 is 4 Thoth. If taken at face value, the date of no. 3 belongs to that portion of the year in which the Babylonian and Egyptian regnal year numbers ought to differ. One expects two year numbers, as is indeed the case in double date no. 1. But no. 3 exhibits only one year number. Would it be the Egyptian number or the Babylonian number? If the regnal Year 9 is Egyptian, then 4 Thoth must be 18 December 457 BC. This is the solution listed in table 1 in section 4.2. However, if the regnal Year 9 is Babylonian, then 4 Thoth must be 18 December 456 BC.

Quite in general, the question arises: If only one regnal year number is given for the portion of the year in which the Babylonian year number and the Egyptian year number differ, then is that one regnal year number Babylonian or Egyptian? I know of only one double date that affords an answer to this question, namely no. 14. In no. 14, the year number is Egyptian. Double date no. 14, namely 20 Adar = 8 Choiak, Year 3 of Artaxerxes II (9 March 402), belongs to the interval in which the Babylonian and Egyptian years do not overlap. Since Artaxerxes II's first Babylonian year began on 10 April 404 and his third on 18 April 402, he could not yet have been in his Babylonian Year 3 on 9 March 402, the date of Kraeling 10.[46] The year numbers therefore ought to differ. But the text provides only a single year number. That year number must therefore be either Babylonian or Egyptian. In the case of no. 14, the year appears to be Egyptian. Double date no. 14 therefore suggests the existence of the following rule: When only a single year number is given and the two year numbers ought to differ, the year number is Egyptian.[47] In fact, a similar rule applies to month and day dates. Dates in Aramaic papyri provide either both the Egyptian and the Babylonian dates or just the Egyptian date, never just the Babylonian date. If the Egyptian date 4 Thoth in no. 3 is not emended, then the rule just stated requires that the single year number be Egyptian. The double date in no. 3 then belongs at the end of 457 BC, and not at the end of 456 BC.

4.3.2. Double Date No. 10. Leads to 17 September 416 BC as Lunar Day 1 and 15 September 416 BC as Closest Conjunction

No. 10 (Kraeling 8) poses a different problem. Lunar Tishri begins on Egyptian 17 Payni, which in Year 8 of Darius II is the same as 17 September 416 BC. The relation of 17 September to conjunction on 15 September 416 BC at about 5:32AM is altogether as expected. But there is a problem elsewhere. In the 19-year lunar cycle observed in Babylon, lunar Tishri began one lunar month later, that is, on or close to 17 October 416 BC. The Babylonian evidence is solid.[48] It has therefore been suggested to emend 22 Payni (Month 10) into 22 Epiphi (Month 11) in double date no. 10.[49] Day 1 of lunar Kislev then corresponds to 17 Epiphi. 17 Epiphi is 17 October in 416 BC. The closest conjunction is on October 14 at about 5:09PM. By this scenario involving emendation, the interval of three days between lunar Day 1 and the day of conjunction is larger than the norm. Emending Payni into Epiphi is therefore less than fully satisfactory.

The alternative is to assume that the text is correct and that the intercalary months were assigned different positions at Assuan from those at at Babylon. Assuan is far away from Babylon.

Twelve lunar months are about 354 days long, about 11 days less than a year. An intercalary thirteenth lunar month therefore needs to be added every two to three years. Most of what we know about the position of intercalary thirteenth months at Babylon derives from the highly regulated system of Babylonian astronomy. It seems eminently possible that this system was not in use wherever Babylonian month names were used. The possibility of variation is rarely discussed because there is hardly any evidence. From lunar dates attested outside Babylon, one cannot normally infer where exactly the lunar month fell. The Aramaic double dates from Egypt are an exception. The position of the lunar months can be inferred from the Egyptian date. The time seems right to press the case that variation in the position of intercalary months was altogether possible in the Achaemenid empire.

In fact, in one other double date, assuming a different position for intercalary months produces a satisfactory match. The double date is found in Papyrus Egyptian Museum Cairo no. 3652 = JdÉ 37114.[50] The date has not been included in the list above because the text is slightly fragmentary. Of the number 21 in Mesore 21, only the 1 is clearly visible. But there are traces of what may well be 20.[51] The resulting double date is as follows.

21 Mesore = 21 Kislev of Year 6 of Artaxerxes (I)

21 Mesore in Year 6 of Artaxerxes I corresponds to 1 December 459 BC. If 21 Kislev is 21 Mesore, then 1 Kislev is 1 Mesore, or 11 November 459 BC. The closest conjunction is on 9 November 459 BC at about 3:29AM. The interval between conjunction on 9 November and the beginning of the lunar month on 11 November is well within the norm.

At Babylon, Kislev began no doubt one lunar month later, on or close to 10 December 459 BC.[52] It could be assumed that "Kislev" (Month 9) is an error in the text for "Heshvan" (Month 8).[53] However, Papyrus Egyptian Museum Cairo no. 3467 (= JdÉ 37106)[54] is a fragmentary copy of the same text. The month name Kislev is clearly visible. It seems improbable that the scribe would have made the same mistake in both copies.[55] The text at hand therefore strongly favours the notion that intercalary months can occupy different positions in the year in different parts of the Achaemenid empire.

4.4. Statistical Considerations about the Sample of Aramaic Double Dates

In the beginning, there are two Assumptions. The first assumption is that any Egyptian date in an Aramaic double date can be converted into a date in our calendar owing to the perfect consistency of the Egyptian year's wandering motion. The second assumption is that the Babylonian date is lunar.

By the first assumption, the Egyptian date of a double date is converted into a date in our calendar. That date in our calendar is naturally the same for the Babylonian date of the double date. From the date in our calendar for the Babylonian date, the date in our calendar for Day 1 of the Babylonian month can easily be derived.

By the second assumption, Babylonian Day 1 ought to fall close to lunar conjunction. The approximate times of conjunction are known by computation.

It appears that Babylonian Day 1 does in fact as a rule fall close to conjunction. Could this be a coincidence? Two factors make it extremely unlikely that this match is purely coincidental. The first factor involves three properties of the double dates in the sample. The second factor concerns statistical considerations.

As regards factor one, the sample of double dates exhibits three properties. First, the double dates have been chosen by an objective criterion. Only dates that are completely preserved have been selected. In no case does the reading of the text pose a problem. Second, each date is found in a different document. None of the dates therefore depends on any other. Each date is fully independent from any other date. Third, the double dates in the sample span the entire period from which Aramaic double dates are known, including four different reigns, those of Xerxes I, Artaxerxes I, Darius II, and Artaxerxes II.

As regards factor two, a numerical analysis reveals that the probability of coincidence is extremely low. In other words, it is extremely probable that the two assumptions stated at the outset of the present section 4.4 are correct. Professional statisticians could probably handle the statistical aspect more adequately than is done here. Still, I believe the following considerations to be valid.

A first consideration concerns the degree of agreement of the double dates among one another. 2 dates exhibit 1 agreement. 3 dates exhibit 3 agreements (2 + 1), that is, a first agreement between dates 1 and 2, a second between dates 2 and 3, and a third between dates 1 and 3. Likewise, 4 dates exhibit 6 agreements (3 + 2 + 1), and so on. 10 dates exhibit 45 agreements (9 + 8 + . . . + 2 + 1) and 14 dates exhibit 91 agreements (13 + 12 + . . . + 2 + 1).

Another way of formulating statistical probability is as follows. From the Egyptian date, lunar Day 1 of the Babylonian month is obtained. What is the chance that lunar Day 1 falls close to conjunction? There are 29 to 30 days in a lunar month. If we consider a period of three days close to conjunction, there is about one chance in 10 that Day 1 falls in that three-day period (30 divided by 3). If we consider a period of two days, there is about one chance in 15 (30 divided by 2). There are 14 double dates. 12 of them fall in a period of three days close to conjunction. 11 fall in a period of two days close to conjunction. Suppose we play it safe and accept only 10 double dates and the agreements between them as valid. Also suppose we consider a three-day period around conjunction. The chance that Day 1 falls in that three-day period is one in 10. What is the probability that 10 instances of Day 1 fall in that three-day period? This is like throwing 10 dice that each have 10 sides. There are 36 possibilities for throwing two traditional dice, that is, 6^2. 6 is the number of sides on each dice. 2 is the number of dice. Accordingly, for 10 dice of 10 sides, the corresponding number is 10^{10}, or 10 billion. The chance of all dice falling on a same given side is one in 10,000,000,000. One in 10 billion is also the probability that the agreement between the 10 double dates is pure coincidence. In 11 double dates, Babylonian Day 1 falls in the same two-day period in relation to conjunction. The number of ways in which 11 dates can fall in a two-day segment of a lunar month divided into successive two-day segments is 15^{11} or 8,649,755,859,375, in fact slightly less because lunar months are on average only 29.53 days long. Still, the probability that the dates of the 11 instances of Babylonian Day 1 is coincidence is about one in eight trillion. In other words, if the two assumptions stated above are correct, then the 11 instances of Babylonian Day 1 need to be dated in one of at least eight trillion possible ways. And this happens to be the case. Surely, this is no coincidence.

5. FIFTH MILESTONE in the History of the 365-Day Calendar (1500–500 BC): Why Is Egyptian and West Asian Chronology for 1500–500 BC Sound? The Amarna Connection and the Assyrian Connection

5.1. Sothic Chronology: Not the Ultimate Foundation of Egyptian Chronology of the Third and Second Millennia BC

Everyone agrees that the third and second millennia BC in Egypt are dated by means of so-called Sothic chronology.[56] Sothic chronology is the dating of Egyptian history by manipulating Sothic dates. A Sothic date is a date according to the Egyptian 365-day civil calendar of the first morning rising or reappearance of the star Sirius, named *Spdt* in Egyptian and Sothis in Greco-Egyptian, one early morning in July. The star is invisible for about two months every year. Because the Egyptian civil calendar has no leap year, its year shifts or wanders in relation to the seasons and also in relation to the rising of Sirius, at a rate of about one day in four years. The date of the rising will therefore also shift by a day in about four years, returning to the same point in about 1460 years (365 × 4). Each period of about four years will therefore have its own Egyptian date of the rising. If we know the Egyptian date of the rising, then we can also know the four-year period in years BC in which the rising fell. Importantly, Sothic chronology implies consistency of the wandering year.

There is no denying that two Sothic dates in specific are the basic organizing principle for dating Egyptian history in the second millennium BC.[57]

The first Sothic date is found in a papyrus from Illahun and falls in the reign of a pharaoh who must be Sesostris III of Dynasty 12. The second Sothic date is found on the verso of the medical Papyrus Ebers and falls in the reign of pharaoh Amenhotep I of Dynasty 18. The first Sothic date in all probability falls sometime in 1900–1850 BC. The second in all probability falls sometime in 1550–1500 BC. The details are not relevant to the present argument.

To falsify standard Egyptian chronology, the first step is not to search for flaws in Sothic chronology. It is true that Sothic chronology is an *organizing principle* for dating Egyptian history. However, the veracity of Sothic chronology is itself *in need of outside confirmation*. Such outside confirmation does in fact exist. Sothic chronology is *not* the *ultimate foundation* for dating Egyptian history. The ultimate foundation is the outside evidence that cements the veracity of Sothic chronology and therefore also of the wandering year's consistency. Using Sothic chronology to date Egyptian history is one thing. Establishing the veracity of Sothic chronology is another. It is the outside evidence for Sothic chronology that is the true foundation of standard Egyptian chronology. That outside evidence is the last line of defence of the standard chronology. The concern of the present paper is with the time-period 1500–500 BC. Two pillars are the ultimate foundation of the chronology of that period. They are described in section 5.2.

There is an additional twist, however. Outside evidence can only confirm Sothic chronology to within a certain margin of error, up to as much as 100 years by some estimates, owing to issues of interpretation of the Assyrian sources. Once the probability that Sothic chronology is valid has been accepted within such a margin of error, it is necessary *to return to Sothic chronology* for fine-tuning. Indeed, there are different ways of evaluating Sothic chronology once it is accepted as probable on the basis of outside evidence. According to these different manners of interpretation, the margin of error in dates for the New Kingdom obtained from Sothic chronology can be as much as about 20 years. For the Middle Kingdom, the margin of error can be as much as about 50 years.

5.2. The Two Pillars of Sothic Chronology in 1500–500 BC: Radiocarbon Dating and Assyrian Chronology

The first pillar consists of a match between dates obtained from Sothic chronology and dates obtained independently from Assyrian chronology. The second pillar is radiocarbon dating, which will not be discussed at length here. Radiocarbon dating is only approximate, is still being refined, and is not without its doubters. Then again, radiocarbon dating could make a difference in evaluating alternative chronological models that differ from the standard model by centuries. My impression is that, in the decades ahead, more can be expected from scientific methods used in laboratories for strengthening the standard chronology. This more presumably includes methods that have not yet been discovered.

What is the Assyrian evidence that supports Sothic chronology? It has rightly been observed that, from 1500 BC down to the seventh century BC, Assyria's chronology is the most firmly dated of the entire ancient world.[58] This circumstance is owed to a variety of cuneiform sources including chronicles, king-lists, and lists of eponymous officials (that is, officials whose name is given to the calendar year).[59] From the seventh century BC onward, Babylon's chronology rises to the same standard as Assyria's, and Egypt's to an even higher one.

The focus is on two sets of dates obtained independently from one another. One set consists of dates derived from Sothic chronology for New Kingdom pharaohs (traditionally *ca.* 1500–1000 BC). The other set comprises dates derived from Assyrian sources for Assyrian kings. Both sets are only approximate, owing to a number of factors. But the range of variation does not amount to much more than a couple of decades. According to Sothic chronology and Assyrian chronology, certain pharaohs ruled around the same time as certain Assyrian kings. But can it be confirmed that the rulers in question are indeed contemporary? The need is for independent confirmation. If such confirmation is indeed possible, then one might feel entitled to conclude that Sothic chronology and Assyrian chronology are in all probability both true. And if Sothic chronology is true, then the assumption of the wandering year's consistency is also true. It would be more difficult to assume that Sothic chronology and Assyrian chronology are both wrong but nevertheless match because the two independent errors by a miraculous coincidence just happen to produce the same deviation from the standard chronology.

5.3. The Amarna Connection

Sothic chronology and Assyrian chronology make certain Egyptian rulers contemporary with certain Assyrian rulers. Their contemporaneity would of course be false if either Sothic chronology or Assyrian chronology or both are flawed. Is there independent confirmation that they are contemporaneous? Such confirmation in fact came to light at the end of the nineteenth century in the plane on the east bank of the Nile called el-Amarna and located about 190 miles south of Cairo. The Amarna plane is the site of the city that was Egypt's capital for a couple of decades in the New Kingdom under the famously heretic king Akhenaten. Called Akhetaten in Egyptian, the city was built from scratch and then abandoned a couple of decades later at the end of the so-called Amarna period, which is traditionally dated to the middle of the fourteenth century BC.

In the 1880s, native Egyptians discovered a trove of more than 300 clay tablets inscribed in the cuneiform script in the ancient ruins of Akhetaten. The tablets preserve royal correspondence between Egyptian pharaohs and rulers of West Asia.[60] There is nothing quite like the Amarna tablets anywhere in the collective evidence from the ancient world. They are among the most precious historical evidence ever unearthed. Two circumstances greatly enhance their historical value. First, the tablets establish a direct link to ancient Amarna. The tablets were untouched for centuries, from ancient burial to modern discovery. No interpretative interventions by later historians can have affected or distorted the information that they provide. Second, as royal correspondence, the texts must have been composed and the tablets inscribed with great care by elite scribes, greatly enhancing trust in their reliability.

Amarna was capital of Egypt for only a couple of decades. Any pharaoh, whether as recipient or sender of letters and whether named or not, must be Akhenaten or a close contemporary. What matters therefore for the present argument is the date of the Amarna period as a whole. The period is so short that there is no need to be concerned all that much with the identities of individual pharaohs.

I suspect that additional evidence both from Amarna letters and other cuneiform sources could probably be carefully construed in favour of the standard chronology.

5.4. Synchronisms between Amarna and Mesopotamia

5.4.0. The Synchronism: Definitions

What Is a Synchronism? A synchronism is *a historical fact or cluster of facts that proves that two historical items are contemporaneous.* Synchronisms can assume many different shapes and forms. For example, two people may be described in a letter as doing business with one another. This description itself is a synchronism. The synchronism indicates that the two people, as items of history, lived around the same time. Or a certain tool may be depicted on a temple. This depiction too is a synchronism. The synchronism indicates that the tool was in use at the time the temple was built. Synchronisms involve many considerations that only add to their complexity. For example, in the case of the object depicted on the temple, it would need to be established that the depiction of the object was not added at a later date to the temple. Each synchronism is a case in its own right. Simultaneity by itself is merely the concept of belonging to the same time-period. Synchronisms are facts or circumstances that prove simultaneity.

Direct Synchronism and Indirect Synchronism. A direct synchronism associates two historical entities directly with another with regard to time. An indirect synchronism does the same indirectly. Consider three historical entities A, B, and C. A first direct synchronism may associate A directly with B. A second direct synchronism may associate B directly with C. The result is an *indirect* synchronism between A and C. After all, if A happens at the same time as B and B happens at the same time as C, then A happens at the same time as C.

Synchronisms as Proof of the Correctness of Our Understanding of Calendrical Systems. Consider two independent calendrical systems I and II. Suppose that a historical entity A is dated by system I and a historical entity B is dated independently from A by system II. Then assume that it appears that A and B date to the same time-period. That still does not mean that A and B are simultaneous. The calendrical systems may be misunderstood. The simultaneity of A and B would then be deceptive. However, a synchronism may overcome this obstacle and cement the simultaneity of A and B. For instance, A and B may be mentioned in the same document. Such a synchronism between A and B proves that the simultaneity of A and B is correct. But the synchronism also proves that the two chronological systems that independently date A and B are correctly understood. Indeed, one or two incorrectly understood systems will hardly yield a correct result. Therefore, if a synchronism confirms a simultaneity, then the calendrical systems that produce the simultaneity must be correctly understood.

5.4.1. A Direct Synchronism Involving the Assyrian King Assur-uballit I and Amarna

Amarna Letter no. 16, inscribed on tablet Cairo C 4746, is a letter from a certain Assur-uballit, an Assyrian ruler. Two Assyrian kings of this name are known. According to the Khorsabad king-list,[61] Assur-uballit II had a short reign in the seventh century BC. Assur-uballit I reigned several centuries before Assur-uballit II for 36 years. Of these two rulers, only Assur-uballit I can be the one writing to Amarna. The Khorsabad list details the length of all its reigns in years. The numbers of the reigns can simply be added up as one moves back into the past. By the inner counting of the Khorsabad list, Assur-uballit I can be dated to roughly 1360–1330 BC. He was apparently the first to claim the title "king of the land of Assyria" in his monuments.[62]

The name of Letter no. 16's royal addressee is fragmentary. Everyone except Edmund I. Gordon[63] has read the name as *Na-ap-ḫu-ri-i-ia* (*Napḫuriya*), which is part of Akhenaten's nomenclature. However, as was noted above, the pharaoh's identity does not really matter. It suffices to consider the Amarna period as a whole. According to Sothic chronology, the Amarna period falls roughly around 1350 BC, with a margin of error of hardly more than 20 years.

Assur-uballit I and the Amarna period are dated to the same time by independent chronological systems, Assur-uballit I by means of Assyrian chronology and the Amarna period by means of Sothic chronology. But are the two in fact simultaneous? The need is for a synchronism that seals the simultaneity. The fact that Assur-uballit I wrote to Amarna is such a synchronism. This synchronism also strongly suggests that Sothic chronology and Assyrian chronology are on the whole correctly understood. If either were misunderstood, they would hardly yield the simultaneity of Assur-uballit I and Amarna that is supported by the synchronism that consists in the fact that this Assyrian king wrote to the royal court in Amarna.

The synchronism between Assur-uballit I and Amarna can be strengthened further. Assur-uballit I mentions in the same letter, no. 16, that his "father" Assur-nadin-ahhe had also corresponded with Egyptian kings. Assur-uballit I is no. 73 in the conventional count of the kings in the Assyrian king-list. Kings named Assur-nadin-ahhe do indeed appear slightly earlier in the list, in positions 66 and 71, and nowhere else.

5.4.2. Indirect and Partial Synchronisms Involving Four Babylonian Kings, Karaindash, Kadashman-Enlil, Kurigalzu, and Burnaburiash

The synchronism just discussed in section 5.4.1 involves an Assyrian king. It may well be the most convincing of all the synchronisms involving the Amarna correspondence. That comes as no surprise. Assyrian chronology is the most solid of all the chronologies of the Ancient Near East for the time period in question. However, only two Amarna letters name Assyrian rulers, namely nos. 15 and 16. Both mention a certain Assur-uballit. By contrast, many more letters concern Babylonian rulers, fourteen in fact, nos. 1–14.

Four Babylonian kings, to be exact, are named in Amarna tablets, to wit: Karaindash (no. 10); Kadashman-Enlil (nos. 1–3 and 5); Kurigalzu (nos. 9 and 11); and Burnaburiash (nos. 6–12 and 14). All four are known from other cuneiform sources. It would be useful to exploit also these Babylonian connections for the purpose of validating Egyptian Sothic chronology and by implication the wandering year's consistency. The problem is that Babylonian chronology is insecure for the period at hand. Babylonian kings of the period in question can be dated only to the extent that they can be connected to Egyptian or Assyrian kings.

A synchronism is a tool that is meant to validate that two historical facts dated by independent means to the same time did indeed happen at the same time. Thus, if Amarna and an Assyrian king are dated to the same time independently, the fact that the Assyrian king wrote to Amarna is a *synchronism* that validates the *simultaneity* of two independently obtained dates. It is crucial that the two dates are obtained independently from one another.

In this respect, using the fact that a Babylonian king wrote to Amarna as a synchronism may at first sight seem to involve circular reasoning, as follows. The synchronism is meant to prove that Amarna and the Babylonian king do indeed belong in the same time period as the independently obtained dates suggest. However, the problem is that Babylonian chronology in the time period in question is uncertain. As a result, the Babylonian kings of that time can as a rule be dated only if they can be connected with Egyptian or Assyrian kings, who are securely dated. Therefore, the fact that a Babylonian king wrote to Amarna cannot be used as a synchronism if that fact has already been used to obtain a date for the Babylonian king. Using the synchronism in this way would be circular reasoning. Circular reasoning is when one uses a certain notion as evidence to prove a proposition in which this notion is already included and assumed to be true. A synchronism between two events cannot involve a date of a Babylonian king if that date has already been obtained by means of the very synchronism.

However, in my opinion, circularity of reasoning can be circumvented with regard to the Babylonian data from Amarna. Sothic chronology and the wandering year's consistency can as a result be further strengthened.

Circular reasoning can be circumvented because the four Babylonian kings mentioned in Amarna tablets can be dated *independently* from Amarna, namely by being linked instead to Assyrian rulers. The Assyrian rulers are themselves already dated independently. Three observations apply.

First, in the Synchronistic History,[64] the Babylonians Karaindash and Burnaburiash are mentioned *before*—and *not long before*—the Assyrian kings Enlil-nirari and Adad-nirari I, nos. 74 and 76 in the Assyrian king-list. These two Assyrian kings can both be dated independently *to around 1300 BC*. Second, in the Chronicle P,[65] the Babylonian Karaindash is mentioned *not all that long before* the Assyrian Adad-nirari, who is dated independently *to around 1300 BC*. Third, in the Synchronistic History,[66] the Babylonian Karaindash and the Assyrian Assur-bel-nisheshu, who is independently dated *to the end of the fifteenth century BC*, are listed as contemporaries.

It may be concluded from these three observations that the four Babylonian kings mentioned in Amarna tablets are dated *independently from Amarna to not all that long before 1300 BC*. At the same time, the Amarna tablets are dated to the same time-period independently by means of Sothic chronology. In sum, both Amarna and the four Babylonian kings are dated independently from one another to not long before 1300 BC. The synchronism that validates this simultaneity is the fact that these Babylonian kings are mentioned in Amarna correspondence.

This synchronism is *indirect* and *partial*. But it is nevertheless fully valid. The synchronism is *indirect* because it relies on first linking Babylonian kings to Assyrian kings in order to establish the independence of the dating of the Babylonian kings. It is *partial* because it involves a date *before* which (*ante quem*), and not *at* which, historical events occurred. Circular reasoning is circumvented because the dating of the Babylonian kings does not depend on Amarna. While the synchronism at hand is only partial, it does point in the right direction and adds its own limited degree of confidence to the standard chronology, especially in relation to rival chronologies that assign dates to the same events that deviate by centuries.

5.4.3. A Transitive Synchronism from Amarna Involving the Babylonian King Burnaburiash and Supporting the Synchronism Involving Assur-uballit I (see 5.4.1 above)

More can be extracted from the Babylonian data from Amarna than just this partial synchronism. The data can be used to buttress the other synchronism between Amarna and

Assur-uballit I already described mentioned above in 5.4.1. The Babylonian data from Amarna cement the synchronism by strongly suggesting that the Amarna Assur-uballit is indeed Assur-uballit I. Unlike Assyrian chronology, Babylonian chronology is insecure in the late second millennium BC. One would therefore at first sight not expect Babylonian chronology to play a role in verifying the standard chronology. However, Babylonian kings can serve as a *link* between Egypt and Assyria, regardless of when they reigned. Equation is after all a transitive property. If A(ssyria) is contemporary with B(abylon) and B is contemporary with E(gypt), then A must be contemporary with E. B's function is fully exhausted in linking A and E. The resulting synchronism may be called transitive. One Babylonian king who serves as a link between Assur-uballit I and Amarna is a certain Burnaburiash, as follows.

First A(ssyria) and B(abylonia). In the Synchronistic History,[67] an Assyrian king Assur-uballit and a Babylonian king Burnaburiash are presented as roughly contemporary. The Assyrian king must be Assur-uballit I, who ruled around the mid-fourteenth century BC. The reason is that the Assur-uballit in question is mentioned immediately before Enlil-nirari in the Synchronistic History. When one looks for an Assur-uballit immediately preceding Enlil-nirari in the Assyrian king-list, one finds Assur-uballit I, who is no. 73. Enlil-nirari is no. 74.

Second B(abylonia) and E(gypt). No less than eight Amarna letters (nos. 6–12 and 14) mention a certain Babylonian king Burnaburiash. The same person is concerned in each of these eight letters.

It appears, then, that a certain Babylonian king Burnaburiash is contemporaneous with Assur-uballit I, on the one hand, and with Amarna, on the other hand. If this Burnaburiash is one and the same person, then this Babylonian king is the key link in a transitive synchronism between Assur-uballit I and Amarna. This synchronism also confirms the other synchronism between Assur-uballit I and Amarna already introduced in 5.4.1.

The Amarna Burnaburiash and the Burnaburiash who is a contemporary of Assur-uballit I would indeed appear to be one and the same person. On the one hand, the Amarna Burnaburiash follows within one or two reigns after the Babylonian king Karaindash, as can be inferred from Amarna letter no. 10. On the other hand, the Burnaburiash who is contemporaneous with Assur-uballit I in the Synchronistic History also follows with one or two reigns of Karaindash, as can be inferred from that same History.

A careful search would probably produce additional source material pertaining to Kurigalzu and Kadashman-Enlil that could be manipulated in similar ways to further strengthen the case for Sothic chronology.

5.5. Refutations of the Standard Egyptian Chronology, Including of Sothic Chronology and of the Wandering Year's Consistency

The chronological model that guides and steers all the dates of all the events of ancient history is inductive. The model has exhibited some ability to resist skepticism. All the textbooks follow it within certain narrow margins of approximation. This reigning model has otherwise not gone uncontested. For instance, a team of historians united around Peter James has presented an alternative model.[68] And there are other such models. James' model stands out because it has received much publicity. The single most characteristic feature of James' model concerns the timing of the historical events that occurred in Egypt between the end of the Nineteenth Dynasty and the beginning of the Twenty-fifth Dynasty—and anywhere else in the ancient world at the same time. The standard model assigns close to 500 years to the Egyptian Dynasties Twenty to Twenty-four, from shortly after 1200 BC to shortly before 700

BC. In James' model, these events amounting to five centuries in the standard model are contracted into two and a half centuries. This contraction is inspired by a sense that there is not enough historical evidence from Dynasties 20–24 to fill 500 years. The end of the Nineteenth Dynasty thus drops down, as it were, from about 1200 BC to about 950 BC. The great Ramses II's reign falls around 1250 BC in the standard model. In the alternative model, that date is lowered to about 1000 BC. This compression of dynasties affects above all the standard chronology's three centuries from about 1050 BC to about 750–700 BC. The revised model makes Dynasty 21 contemporaneous to Dynasty 22. Dynasty 22 is greatly reduced in length. Dynasty 22 and the shorter Dynasty 23 in fact already fully overlap in the standard model.

The standard Egyptian chronology of the third and second millennia BC is based squarely on Sothic chronology. And Sothic chronology implies consistency of the wandering year. James' model therefore necessarily involves the rejection of Sothic chronology and of the wandering year's consistency.

I do not think there is anything that I could write to persuade James or defenders of other alternative theories of the veracity of the standard chronology. They seem comfortable with their positions. And so am I with my confidence in the standard model. No new dramatic evidence that I am aware of has come to light since James' *Centuries of Darkness* (1991/1993). Still, the publicity and attention that rival chronologies garner underline the need for revisiting the reasons why all the handbooks of history share a single chronology back to 1500 BC.

Having been familiar with the relevant data for a while now and looking ahead, I am convinced that the overall dating of ancient history will never substantially change. Alternative models have not made one dent in the chronology found in the textbooks. In my estimation, the standard model will prevail in all perpetuity by presenting a more coherent solution for the data relevant to ancient chronology than anything any skeptics exploiting individual points of uncertainty may be able to produce. Absolute certainty is too much to hope for. But I have no worry. My understanding of the sources has convinced me personally that the standard chronology will keep successfully resisting the inevitable skepticism.

Nor do I believe that the reigning model would persevere out of sheer inertia, let alone by a conspiracy of academics who do not want the peace disturbed. One might remember that, until half a century ago, the Great Pyramid at Giza was often dated to much earlier than the now commonly accepted 2500 BC. A radical revision of ancient chronology would be welcome. Attempts should be encouraged. But I do not think that any such attempt will ever be successful.

No one doubts that the sources are replete with errors and contradictions. If complete absence of errors in the sources is a prerequisite for establishing a chronology of antiquity, no such chronology will ever be possible. If all it takes to falsify a chronological model is to point to a contradiction in a text, then nothing is easier than to falsify ancient chronology in general and Sothic chronology in specific.

Challenges to the standard model evoke an interesting facet of the field of Egyptology. If one were to ask Egyptologists point blank how we know that the great Ramses II reigned roughly in the thirteenth century BC, one might find the possession of the answer to this simple and basic question to be rather rare. Chronology is a specialized subject. Not everyone feels called to engage it. Then again, BC historians can much less afford disengagement from chronology than AD historians.

A certain prevalent unawareness of the very foundations of ancient chronology makes it wiser at this time to devote more effort to promoting understanding of the standard model than to warding off critiques of it. Ideally, as many ancient historians as possible would be in the possession of the requisite analytical tools to make up their own mind completely independently. This paper is meant to inform rather than to persuade.

Two characteristics of James' model facilitate an assessment of his views. First, the departure that he proposes from the standard chronology is localized. It does not involve a wide range of revisions affecting many periods. It is a one-point theory. The big point is that there is not enough evidence to fill five centuries in the standard model's 1200–700 BC. The absence of sufficient evidence is interpreted metaphorically as darkness, hence the book's title *Centuries of Darkness*. As quantities of evidence goes, it is not clear to me how many tombs, settlements, statues, and so on, are needed to fill a century or how exactly one measures that. There is no denying that, in the period in question, the ancient Near East witnessed a pronounced absence of strong unified states. This absence can be interpreted positively as decentralization or negatively as chaos. At any event, sequences of rulers become more difficult to follow and the chronological framework turns more opaque. James also makes much of perceived similarities between artistic styles found in various parts of the ancient world and now dated to different time-periods. Similarity would seem to suggest simultaneity. Then again, artistic style is not the most objective of criteria and dating style features requires caution.

The second characteristic of James' revision is that the centuries on which it focuses are relatively recent in antiquity. The contested period does not fall around 2500 BC or around 1500 BC—that is, 4500 or 3500 years ago—but around 1000 BC. These are the centuries just before the time when the great civilizations of Greece and Rome emerge prominently on the world's historical scene and when the chronology of the ancient world becomes a matter of universal consensus, that is, from 700/500 BC onward. Even James does not contest the traditional chronology of the ancient world from about the seventh century BC onward. Obviously, the farther into the past one recedes, the scarcer the historical evidence becomes. Thus, the evidence (and hence the web of chronology) is much denser after the beginning of the Egyptian New Kingdom, traditionally dated to around 1500 BC, than before. The evidence includes more plentiful information about relations between the nations of the ancient Near East. As a result, confrontations between James' alternative model and the standard model are bound to be better nourished by hard evidence than if the contested period had occurred earlier in antiquity. The contested period falls as late as possible in antiquity before relative certainty in matters of chronology sets in.

The synchronism between Assur-uballit I and Amarna described above (5.4.1) appears convincing and is generally accepted as proof of the veracity of the standard chronology. But the evidence from Amarna letter no. 16 fails to impress James, who deems the synchronism "flawed".[69] He points out that Assur-uballit calls Assur-nadin-ahhe "father". But it is king no. 72, Eriba-Adad, who is Assur-uballit I's "father" on the monuments. Most have assumed that Assur-uballit I uses the term "father" in the extended meaning of "ancestor".[70] James finds this assumption unacceptable.

In rejecting this synchronism, James implies that the Amarna Assur-uballit is not Assur-uballit I, the Assyrian king-list's no. 73, dated traditionally to the mid-fourteenth century BC, but rather "another, as yet unattested, ruler".[71]

James' dismissal of the synchronism comes as something of a surprise, and not because it seems difficult to believe that the data listed above could be coincidental or deceptive. Rather, James' dismissal of the synchronism is in danger of exhibiting a certain logical inconsistency. By his own model, the chronologies of the nations in the ancient Near East are all reduced by 200 to 250 years. The reduction applies to both Amarna and Assur-uballit I. If both are reduced in date by the same amount of time, why could they not still be simultaneous in James' model? The synchronism might easily be construed as an argument in favour of James' model. His rejection of the synchronism makes him seem overly zealous in discrediting the standard model.

James also rejects the use of mentions of Babylonian kings in the Amarna letters discussed above as evidence. Such use would be, in James' view, "dangerously circular reasoning".[72] It has been argued above in 5.4.2, however, that no circular reasoning need to be involved.

In James' model, the transitive synchronism involving king Burnaburiash described above in 5.4.3 also plays no role. However, that is because James erroneously states that Kadashman-Enlil, rather than Burnaburiash, is the Babylonian king who refers to Karaindash as a recent ancestor in Amarna letter no. 10. Kadashman-Enlil otherwise does receive mention in letters nos. 1–3 and 5.[73]

Space limitations do not allow a more detailed discussion of characteristics of chronological models that significantly lower the dates of events. For example, in James' model, Dynasty 22 is much reduced in length, from about 220 years to about 150 years. And yet, accepting the traditional length of Dynasty 22 seems effortless at first sight. First of all, James himself notes that "[e]vidence of all types is abundant, indeed extremely rich".[74] Then there are the highest attested regnal years of pharaohs. They indicate the minimum length of their reigns in years. For example, if a Year 10 is attested for a certain pharaoh on a monument, then that pharaoh has at least reigned nine to ten years. If one adds up the highest regnal years of the rulers of Dynasty 22, namely 21, 33, 14, 30, 39, 10, 6, and 38,[75] the total is 221 years, altogether in line with the length of Dynasty 22 according to standard chronology.

There is no room here for introducing new elements into the discussion. One such element is language. Late Egyptian and Demotic are the third and fourth stages of the Egyptian language respectively. The Late Egyptian of the Story of Wenamun, dated to Dynasty 21, is very different from the Egyptian found in the earliest Demotic documents of about 640 BC. Linguists now have to assume that it took 350 to 400 years for the Egyptian language to evolve from the Egyptian of Wenamun to earliest attested Demotic. By James' model, 150 to 200 years would need to suffice. The matter deserves further inquiry.

5.6. How Established is the Chronology of 1500 BC–500 BC?

A comparison with the history of astronomy may serve to answer the question. As recently as 1867, in one of the many editions of his famed *Outlines of Astronomy*, John F. W. Herschel presents the Copernican view of the solar system defensively as an eminently probable result inferred from inductive reasoning. Today, 150 years later, in the space age, few would doubt that the earth is round and that it revolves around the sun, even if there may still be skeptics. Ancient chronology now seems to be in a similar stage of development as astronomy was a century and a half ago. It is not clear yet whether the standard chronology will move toward a state of virtual certainty in the way that our notion of the solar system has. I have difficulty imagining what kind of Egyptian source or what kind of other evidence might emerge that would dismiss all doubt about the standard chronology.

6. SIXTH MILESTONE in the History of the 365-Day Calendar (Early Third Millennium BC): The Origin of the Egyptian 365-Day Calendar

6.1. The Probable Role of the Rising of Sirius

It was suggested above that chronology is a story best told backward by starting at the end. However, beginnings never fail to fascinate. Accordingly, hardly any topic in Egyptian calendrics has been discussed more than the origin of the Egyptian 365-day year. I have elsewhere described the state of the discussion on this problem.[76] Upon further reflection, the problem seems clearer to me now than it did even just a couple years ago with regard to one point in specific. That specific point concerns the role of the rising of Sirius in the origin of the Egyptian civil calendar. I am certainly not the first to associate the beginning of the civil calendar with the rising of Sirius. I cannot positively prove that the very first 365-day year ever, presumably some time in the early third millennium BC, in fact began with a rising of Sirius. Yet, there are so many facts that speak in favour of this assumption that I would now consider it by far the most probable. At the very least, it may be useful to list the facts in favour. As facts, their validity cannot be denied. Let others decide for themselves as to how suggestive their combination is.

As the earth revolves around the sun, the sun with its blinding light positions itself at some point roughly between the earth and the star Sirius, the brightest in the night-sky. As a consequence, the star Sirius is completely invisible for about two months. But as the earth progresses on its orbit around the sun, Sirius becomes visible again one early morning just before sunrise in high summer. That first visibility after weeks of absence is called the rising of Sirius. The Egyptian term is *prt spdt* "the coming out of *spdt* (Greek form: Sothis)".

Undeniable facts can be adduced in favour of the following three tenets:

Tenet (1): The original calendar of ancient Egypt was lunar.
Tenet (2): The 365-day civil calendar is derived from the lunar calendar.
Tenet (3): The original lunar year began around the time of the rising of Sirius.

These three tenets easily suggest, even if they do not prove, the following conclusion:

The 365-day civil calendar, having been derived from the lunar calendar, also began with the rising of Sirius.

The following three facts (a), (b), and (c) favour tenet (1) above, pertaining to the lunar character of the earliest Egyptian calendar:

(a) The hieroglyph used in the word for month, *ꜣbd*, depicts the moon.
(b) The length of the civil month, 30 days, must be derived from the length of the lunar month. The astronomical lunar month is about 29.53 days long. Calendrical lunar months are therefore either 29 days or 30 days.
(c) The word *ꜣbd* "month" is of lunar origin, denoting Day 2 of the lunar month, in the evening of which the first crescent is presumably seen.

In addition, the earliest calendar in most if not all nations of the world appears to be lunar. Why would the Egyptian calendar be different?

The three facts (a), (b), and (c) also favour tenet (2), the derivation of the civil calendar from the lunar calendar. That is because each of these three facts reveals a feature of the civil calendar that points to lunar origin. In addition, one might also consider that most if not all other non-lunar calendars known have been derived from lunar calendars. Their months are all around 30 days long. Why would the Egyptian civil calendar be an exception?

I have elsewhere listed seven facts supporting tenet (3), that is, the notion that the original lunar calendar began with the rising of Sirius.[77]

(1) *First fact.* In pBerlin 10056, A, from Illahun, a time unit called *rnpt* "year", consisting of lunar months, begins around civil II *šmw*.[78]
(2) *Second fact.* In the Illahun archive, the rising of Sirius (*prt spdt*) falls generally in late IV *prt*.[79]
(3) *Third fact.* In the Ebers calendar, a set of 12 names pertaining in one way or another to months begins with the rising of Sirius.
(4) *Fourth fact.* In the Canopus Decree of 238 BC, *prt spdt* is explicitly equated with *wp rnpt*. Other sources point to the same equation.[80]
(5) *Fifth fact.* At Illahun and elsewhere, the civil *w3g*-feast falls on I *3ḫt* 18, that is, at the very beginning of a year or of a set of months.
(6) *Sixth fact.* In the same Illahun archive, by contrast to the civil *w3g*, the non-civil *w3g* falls generally in II or III *šmw*.
(7) *Seventh fact.* pBerlin 10007 from Illahun exhibits the following sequence: *prt spdt* falling in IV *prt* (line 17) – *w3g* (line 18) – *wp rnpt* (line 22) – *w3*[*g*] (line 23). This suggests a lunar year beginning in IV *prt* with the rising of Sirius followed by its own lunar *w3g*, and later by the civil new year (*wp rnpt*) followed by the civil *w3g*. The religious festival year tends to be lunar and therefore has precedence in papyri from funerary domains.

One additional consideration at least does not contradict the notion that the very first Egyptian 365-day year began with the rising of Sirius. The standard chronology assumes that the Egyptian year wandered with perfect regularity. According to this model, New Year's Day falls close to the rising of Sirius in 2800/2700 BC, when writing and the 365-day calendar are first attested. This closeness of New Year's Day to the rising of Sirius is fully independent from the other indications that the rising of Sirius is associated with new year. In being independent, the two confirm one another.

6.2. "Inundation" as a Translation of *3ḫt*, the Name of the First Season

Even more conspicuous than the rising of Sirius as a natural phenomenon is the inundation of the Nile. No natural phenomenon is more specific to ancient Egypt's reason for existence as a nation. The flood is often related to the Egyptian civil calendar. The rising of Sirius happens to fall near the beginning of the period in which the Nile rises. It is clear from the sources that the ancient Egyptians associated the two with one another. Therefore, a year that begins with the rising of Sirius is automatically also a year that begins with the onset of the inundation. The undeniable facts listed in 6.1 above make it much more likely that the original lunar calendar began with the rising of Sirius than with the inundation. Still, the inundation may later have come to be viewed as an additional marker of the beginning of the year.

The main focus of the present section is on another calendrical concept, namely the universal assumption that *ꜣḫt* as name of the first season of four months means "inundation". Upon closer inspection, this assumption is far from secure. It appears again that the foundations of Egyptian chronology require constant questioning.

The most common assumption about the connection between the flood and the civil calendar is that *ꜣḫt*, the name of the first season, means "inundation". By now, this assumption is so deeply entrenched that it may seem like heresy to doubt it. Yet, the translation "inundation" does not seem warranted by anything that is known about roots such as *ꜣḫ*, or *ꜣḫ*(*y*), or *jꜣḫ*(*y*). A more systematic investigation is desirable. Sethe, who collaborated on the Berlin *Wörterbuch*, admitted decades ago that the evidence for "to flood" as a translation of *jꜣḫ*(*y*) is sparse. He notes that any potential evidence is limited to the Pyramid texts, whose interpretation is often difficult.[81] A cursory survey suggests that translating *jꜣḫy* as "be inundated" is not binding in the Pyramid texts. Related translations such as "be verdant" or "flourish" seem equally possible. The other two season names, *prt* and *šmw*, later came to form a contrastive pair denoting the opposition between the cold season or winter and the hot season or summer. The Coptic forms are **ⲡⲣⲱ** "winter" and **ϣⲱⲙ** "summer".

"Inundation" was in fact not the normal translation of *ꜣḫt* in earliest Egyptology. Both Jean François Champollion (1790–1832) and Karl Richard Lepsius (1810–1884) translate another season name, *šmw*, as "inundation".[82] Champollion interpreted the hieroglyph in *ꜣḫt* as a garden with vegetation and translated *ꜣḫt* accordingly. In his opinion, *ꜣḫt* "vegetation" naturally follows *šmw* "inundation". Champollion read the second season as *hr*(*t*) (**ϩⲣ**) instead of *prt*, still confusing (*h*) and (*pr*).

Champollion's essay on the calendar appeared in 1842 and Lepsius' foundational treatise on chronology in 1849. But the prevailing opinion on the meaning of *ꜣḫt* had changed by 1856, when Heinrich Karl Brugsch (1827–1894) published his *Nouvelles recherches sur la division de l'année des anciens Égyptiens*.[83] Brugsch vigorously defended "inundation" as a translation of *ꜣḫt* and not of *šmw*, against adherents of Champollion's view such as Jean-Baptiste Biot (1774–1862) and Emmanuel vicomte de Rougé (1811–1872). Brugsch returns to the matter in both his *Matériaux*[84] and his *Thesaurus*.[85] The theory that *šmw* means "inundation" was therefore fairly short-lived in Egyptology.

"Inundation" is now universally accepted as a translation of *ꜣḫt*. But a preliminary investigation reveals that the arguments in favour of this translation are not strong. Some arguments are even based on what have been shown to be false assumptions or obsolete interpretations of texts or other evidence. Nor does it help that Brugsch is known, not only as a giant in the field, but also as someone often given to unfettered speculation. A detailed archeology of Brugsch's exuberant mind, as reflected in his writings, would be necessary to document the very beginnings of the theory that *ꜣḫt* means "inundation".

Two examples of arguments in favour of "inundation" as a translation of *ꜣḫt* are as follows. First, Brugsch adduces a text inscribed on the late temples of Edfu and Dendera that places *bꜥḥ* "inundation" in the season *ꜣḫt*.[86] But can one rightfully conclude from this association that *ꜣḫt* means "inundation"?[87]

Second, Sethe derives the translation of *ꜣḫt* as "inundation" directly from the existence of a Sirius year beginning with the rising of Sirius in July.[88] In such a year, the first season *ꜣḫt* would be "inundation". However, there is no evidence for a Sirius year with Sirius seasons, Sirius months, and Sirius days. The supposed Sirius year can therefore hardly be an argument in favour of "inundation" as the translation of *ꜣḫt*.

Whence has "inundation," as a translation of *ꜣḫt*, drawn its strength over the decades? In my opinion, the following implied reasoning has played a role. The reasoning begins with two undeniable facts. First, *ꜣḫt* is the first season. Second, inundation is viewed as a beginning by the ancient Egyptians. The following two equations result from these facts.

ꜣḫt = beginning season
inundation = a beginning

In an unconscious application of the property known in mathematics as transitivity, a third equation has entered the field of Egyptology.

ꜣḫt = inundation

There is no direct evidence whatsoever that supports this third equation. However, the following characteristic of the reigning chronology may, again unconsciously and likewise unjustifiably, be viewed as supporting the third equation.

If one follows the wandering motion of the Egyptian year back to the early third millennium BC, when the civil calendar was presumably invented, it appears that the season *ꜣḫt* coincides or overlaps at that time with the rise of the Nile, which lasts 100 days or so from about June to about September. It is then a small step to assuming that *ꜣḫt*, the first season, was named after the striking event that occurred in it, namely the flood.

However, another reason was suggested above of why the first *ꜣḫt*-season ever began when it did in the early third millennium BC. The season, the first of the year, may well have begun with the rising of Sirius, just like the original lunar calendar. And the morning rising of the sky's brightest star just happens to fall near the beginning of the flood. But that in turn may just be coincidence. To which extent the rising of Sirius was chosen or at least favoured because it coincides with the beginning of the flood we will never know. There is every reason to believe that the flood is secondary as a calendrical principle. The main argument is that the original lunar calendar clearly begins with the rising and not with the flood.

6.3. The Number 365

The Egyptians were the first to use the number 365 for calendrical purposes, preceding all other nations by many centuries.[89] The earliest secure attestation of 365 as the number of the days of the Egyptian year seems to date to the reign of Mycerinus. In the annals on the Palermo-stone, the number 365 can be inferred from how the transition from Mycerinus to Shepseskaf is represented. "Erst beim Regierungswechsel von Mykerinos zu Schepses-kaf (Palermo, R[ück]s[eite] Reihe 2)", writes von Beckerath,[90] "läßt sich der Gebrauch des Kalenders von 365 Tagen nachweisen, wie Borchardt[91] demonstriert hat". In fact, the earliest comprehensive representation of the divisions of the year into seasons and epagomenal days also dated around the time of Mycerinos. The representation is found in an inscription from the tomb of *N.k-ꜥnḫ* at Tehne.[92] Mycerinus is mentioned in the inscription. The epagomenal days are listed in the table before the three seasons.

It remains unresolved how the Egyptians obtained the number 365. There is no historical document describing the invention of the civil calendar. In fact, this invention may well predate the emergence of full-fledged hieroglyphic writing. It must have become obvious at some point to early Egyptians that there are about 30 days in a lunar cycle and that there are about

12 lunar cycles in one cycle of seasons. Combining these two observations automatically produces 360 as a schematic count of the days of the year. That still leaves the origin of the five epagomenal days unresolved.

Several hypotheses have been proposed regarding the discovery of 365. First, one might have counted the days that pass between two successive instances of a certain yearly phenomenon pertaining to a star, such as the rising of Sirius. Second, one might have measured the distances in days between beginnings of the flood every year over a number of years and then take the average. Neugebauer favoured this scenario, but it seems improbable.[93] Third, one can add up lunar years of 12 months and lunar years of 13 months over a number of years and take the average. Parker leaned toward this solution.[94] There is no way of verifying any of these hypotheses.

7. Conclusion

What is said above does not present much that is new. However, a widespread malaise seems to plague the modern study of ancient chronology. There is a perception that ancient chronology has become inaccessible and impenetrable to many if not most because its foundations and structural outline are not often enough or sufficiently clarified and articulated. Consequences of this malaise include a vague sense of diffidence towards the current ancient chronology and general defencelessness on the part of ancient historians against those that would criticize it. One remedy is exercises whose aim is to revisit the true foundations of modern models of ancient chronology. What precedes is such an exercise.

Notes

1. As early as 1861, von Gutschmidt (1889), p. 366, described chronology's role in BC history, especially Egypt's, as follows:

   In der Geschichte Aegyptens, deren epigraphische Urkunden ausser Königsnamen und Datum nur zu häufig nichts Wissenswertes enthalten, deren beste griechische Quelle in einem Königsverzeichniss [that is, Manetho] besteht, in der endlich die Bestimmung des Alters selbst ein so hohes weltgeschichtliches Interesse hat, muss die Chronologie der Natur der Sache nach eine hervorragendere Rolle spielen als in der Geschichte jedes anderen Landes; kein Wunder, dass sie ein besonders beliebter Gegenstand der bisherigen Forschungen gewesen ist.
2. Depuydt (2005). Among other recent work that strengthens the standard ancient chronology from about 700 BC onward is Jonsson (1998) and Young (2004). Jonsson is also preparing a comprehensive survey of the chronology of the sixth and fifth century BC focusing on cuneiform and Aramaic sources. His work on chronology is inspired by a desire to challenge chronologies perpetuated to this day in conservative religious circles in opposition to what all the handbooks of ancient history teach.
3. For full details about the formula's mechanics, see Depuydt (1997a, 1999).
4. Ideler (1806).
5. For a detailed account of the names and of how these names reveal the structure of the calendars and the relations of these calendars to one another, see Depuydt (1997a, 1999).
6. For details, see Depuydt (1997a), pp. 109–136.
7. Champollion (1842).
8. Lepsius (1849), p. 135.
9. Champollion (1842).
10. Salvolini (1833).
11. Lepsius (1849).

12. The following appreciation by Bunsen (1845–1857), vol. IV, p. 32, of Champollion's discovery of the seasonal names is therefore hyperbolic to say the least:

    Der ersten Abtheilung aber steht der Heros dieses Buches voran. Groß wie Champollions Ruhm und sein Verdienst um die ägyptische Geschichte ist, deren Denkmälern er zuerst ihre Sprache widergegeben hat, so muss doch die von ihm auf dem Gebiete der astronomischen Gleichzeitigkeiten gemachte Entdeckung und die geniale Anschauung eines Allen offenen, einfachen Umstandes, auf welcher jene Entdeckung ruht, dem Historiker das Wichtigste und Glorreichste heissen. Die Distichen versuchen von der Idee dieser Entdeckung ein anschauliches Bild zu geben.
13. Ideler (1825–1826), vol. 1, pp. 94–99.
14. For the exact day and time of day, namely late afternoon of 11 June 323 BC, see Depuydt (1997b); see also, similarly and independently (11 June 323 BC), Walker (1997), p. 25. Instead of Walker's ((1997), 25, top) 13 Ululu as the date of the eclipse that occurred about 10 days before the battle of Gaugamela (see tablet BM 36761 + 36390), I suggest 14 Ululu (Depuydt (1997b), pp. 124–125, note 23).
15. On the Canon, see Depuydt (1995c).
16. In his *Œuvres complètes* 21, at pp. 150, 626, and elsewhere (see Neugebauer (1975), p. 1061, note 1).
17. Ideler (1825–1826), vol. 2, pp. 587–588.
18. Neugebauer (1962), pp. 3–4.
19. Ideler (1825–1826), vol. 1, p. 77.
20. Scaliger (1583) (first edition; Paris), (1598) (second edition; Leiden), (1629) (third edition; Geneva).
21. Crusius (1578).
22. Scaliger (1583).
23. Ideler (1825–1826).
24. Grafton documents two prior efforts to count years back to before AD. The first is a marginal annotation by Erasmus Reinhold in the margin of a copy of his *Prutenicae tabulae coelestium motuum* (Tübingen, 1551), namely British Library C. 113 c. 7 (1), at fol. 23r: *Usus anni Juliani retro etiam ante Christum et Caesarem cogitandus est* "The use of the Julian year can also be thought backward to before Christ and Caesar" (Grafton (1993), p. 120, note 17). The second is Paul Crusius' count of years backward from 1 January AD 33 as an artificial date of Jesus of Nazareth's death or of the Passion of Christian tradition, in his *Liber de epochis* of 1578 (Grafton (1993), p. 133).
25. Neugebauer (1975), vol. 3, p. 1062, note 4.
26. Grafton (1993), p. 203.
27. Rühl (1897), p. 2. Ideler (1825–1826), vol. 2, p. 604 evokes the contrast between Scaliger and Petavius as follows.

    Bei aller Gelehrsamkeit und allem Scharfsinn indessen hat [Scaliger] sich viele bedeutende Irrthümer zu Schulden kommen lassen, die seiner lebhaften, zu Hypothesen geneigten Phantasie und der Beschränktheit seiner astronomischen Kenntnisse zuzuschreiben sind. Auf seine Schultern trat Dionysius Petavius, der mit gleicher Gelehrsamkeit und nicht geringerem Scharfsinn einen ruhigern Prüfungsgeist und einen ungleich größern Vorrath astronomischer Kenntnisse verband.
28. Grafton (1993), p. 110.
29. Grafton (1993), pp. 109–115.
30. Grafton (1993), p. 109.
31. Grafton (1993), pp. 114–115.
32. Grafton (1993), pp. 197–198.
33. *Quarto Calendas Septembris* (for *Septembres*) means "on the fourth day (before) 1 September" (1 September being the first day, 31 August the second day, 30 August the third, and 29 August the "fourth").
34. Swerdlow and Neugebauer (1984), pp. 183–188.
35. Swerdlow and Neugebauer (1984), p. 186.
36. Thus J. Funck, *Chronologia* (Basel, 1554), vol. 2, p. 17, as cited by Grafton ((1993), p. 131).
37. See note 14 above.
38. Grafton (1993), pp. 131–133.

39. Grafton (1993), p. 271.
40. For the transmission of Ptolemy in Arabic sources, see Kunitzsch (1974), as well as other work by this same author. For a recent survey of Byzantine astronomy, see Tihon (1994), who writes (p. 614): "Les Byzantins étaient conscients d'être les détenteurs des connaissances scientifiques de l'Antiquité grecque et Ptolémée était leur astronome". For a recent survey of Arabic astronomy, see Saliba (1994), who emphasizes (p. 28) Copernicus' role as heir of Arabic science.
41. B2, B3, and C3 are the sigla of the manuscripts in Porten (1986–1993). A concordance is as follows (see Porten (1986–1993), vol. 2, p. vii). Cowley is Cowley (1923) and Kraeling is Kraeling (1953).

Edition	Manuscript
1. C3.8 (Porten (1986–1993))	Memphis Shipyard Journal
2. B2.1 = Cowley 5	Bodleian Library (Oxford) MS. Heb. b. 19 (P)
3. B3.1 = Cowley 10	Staatliche Museen zu Berlin P. 13491
4. B3.2 = Kraeling 1+18/4	Brooklyn Museum 47.218.152 + 47.218.155 Frag. 4
5. B2.7 = Cowley 13	Egyptian Museum Cairo 3649 = JdÉ 37108
6. B2.8 = Cowley 14	Egyptian Museum Cairo JdÉ 37112
7. B3.4 = Kraeling 3	Brooklyn Museum 47.218.95
8. B3.5 = Kraeling 4	Brooklyn Museum 47.218.91
9. B3.6 = Kraeling 6	Brooklyn Museum 47.218.32
10. B3.9= Kraeling 8	Brooklyn Museum 47.218.96
11. B2.10 = Cowley 25	Egyptian Museum Cairo JdÉ 37113
12. B2.11 = Cowley 28	Egyptian Museum Cairo 3650 = JdÉ 37109
13. B3.10 = Kraeling 9	Brooklyn Museum 47.218.92
14. B3.11 = Kraeling 10	Brooklyn Museum 47.218.88

In the table of Aramaic double dates in section 4.2 in my chapter Depuydt (2006b), add "Year 8" under "12 Thoth" in date no. 11 and "Year 13" under 9 Hathor in date no. 12 (as pointed out to me by C. O. Jonsson and correctly reported earlier in Depuydt (1995b), p. 162). Both dates have *two* numbers.
42. The times for conjunction used here are those computed by Goldstine (1973). These times are still accurate enough for use by historians. But see now also the times provided by Fred Espenak at http://sunearth.gsfc.nasa.gov/eclipse/phase/phasecat.html.
43. Depuydt (2002), pp. 471–477. See also, more recently, my paper entitled "First Crescent Visibility's Irrelevance: Additional Evidence, including from Babylonian Astronomical Texts," read at the 214th Annual Meeting of the American Oriental Society, which was held in San Diego, California, in March 2004. The thesis of this paper was summarized as follows in the abstract that was published for limited circulation among conference participants:

> Few assumptions of ancient history are as stubborn or have been as dominant from time immemorial as the one that ancient months, which were mostly lunar, began with first crescent visibility. Yet, quite to the contrary, it would appear that first crescent visibility nowhere in the ancient world marked the beginning of the month.

I realize that this is a view that runs against rather strong currents of common opinion and would make much of what is written about first crescent visibility from a calendrical perspective, except with regard to the Muslim religious calendar, into a red herring. In his paper for the Notre Dame session, John Steele showed how the lengths of astronomical months could have been calculated at Babylon from about 350 BC onward. Steele's contribution comes against the background of the still dominant view that Babylonian astronomers had first crescent visibility in mind as a sort of beacon or marker, even if their calculations occasionally resulted in months beginning a little too early to obey the first crescent visibility principle. I personally believe that not even in Babylonian astronomy did first crescent visibility play a role as the beacon of computation, in spite of all kinds of indications to the contrary. I would hope to elaborate on this matter elsewhere in print.
44. For a recent assessment—in a different context—of the limited value of astronomically precise calculations for the evaluation of the lengths of ancient lunar months, see Wells (2001).
45. Porten (1990), p. 25.

46. Cf. Depuydt (1995b), p. 160.
47. Depuydt (1995b), p. 160.
48. Parker and Dubberstein (1956), pp. 9, 33.
49. Porten (1990), pp. 23–24; Horn and Wood (1954), pp. 16–17.
50. No. 8 in Cowley (1923); "B2.3" in Porten (1986–1993).
51. Porten (1990), p. 24.
52. Parker and Dubberstein (1956), pp. 8, 32.
53. Porten (1990), p. 24.
54. No. 9 in Cowley (1923); "B2.4" in Porten (1986–1993).
55. Porten ((1990), p. 24) does consider the possibility of difference in intercalation when he writes as follows:

    Alternatively, we achieve the same date if we assume a failure to intercalate a second Adar in 459. But such a failure would have been strange since three years would already have elapsed since the previous intercalation in 462. Without intercalation 1 Nisan would have fallen on March 20, a uniquely early date.

56. On Sothic chronology in more detail, see Depuydt (1995a, 2000).
57. For a recent survey of attested dates, see Goddio-Von Bomhard (2005).
58. See von Beckerath (1994), p. 19 (cf. Id. (1997), p. 59); on Assyrian chronology, see more recently Postgate (1991) and Hagens (1995) (with additional bibliography). von Beckerath's two books are a treasure-trove of information by a recognized expert on the subject. I have drawn great benefit from them and remain a grateful user. But what is lacking in them is a more articulated exposition of the logical structure of ancient chronology.
59. For the evidence, see Brinkman (1976), Grayson (1975), Millard (1994), and Schmidtke (1952).
60. See Moran (1992) and also Hess (1993). For a large selection of hieroglyphic Egyptian texts from Amarna in translation, see Murnane (1995). The Egyptian texts do not mention Mesopotamian rulers.
61. Schmidtke (1952), pp. 52–56 (as based on the critical editions and notes by Poebel (1942/1943) and Weidner (1944)).
62. Hallo and Simpson (1998), p. 113.
63. Moran (1992), p. 39.
64. Grayson (1975), pp. 158–160; Schmidtke (1952), pp. 84–85.
65. Grayson (1975), pp. 171–175.
66. Grayson (1975), p. 158; Schmidtke (1952), p. 84.
67. Grayson (1975), p. 159; Schmidtke (1952), p. 85.
68. James (1991/1993).
69. James (1991/1993), p. 340.
70. For instance, the use of the term "father" as "ancestor" in colophons to Babylonian texts is discussed in Neugebauer (1955), pp. 13–15 (reference owed to John Steele).
71. James (1991/1993), p. 340.
72. James (1991/1993), p. 343.
73. James (1991/1993), p. 343.
74. James (1991/1993), p. 222.
75. von Beckerath (1997), p. 95.
76. Depuydt (2001).
77. Depuydt (2000), § 6.3.
78. Luft (1992), p. 74.
79. Luft (1992), pp. 156–157.
80. Parker (1950), pp. 33–34.
81. "Das Wort für 'Überschwemmung' . . . ist früh ausgestorben; es findet sich ebenso wie das zugehörige Zeitwort *ỉꜣẖj* nur noch in den alten Pyramidentexten. Die späteren Zeiten verwenden dafür *mḥj* oder *ḥꜣj*" (Sethe (1919–1920, p. 295, note 1). Earlier, in 1900 and 1904, Sethe had devoted two other articles to the season *ꜣẖt*. In 1900, he confirmed that word contains the phoneme *ẖ*, citing a striking example from Niussere's sun temple. Furthermore, the fact that the word pertains to water and lakes in several passages in

the Pyramid texts (p. 106) supports the translation "inundation." In 1904, Sethe provides additional arguments in favour of *3ḫt* as the most probable reading of the word in the form of instances of the word found in inscriptions from Tehne and on hieratic ostrakon Leiden J429 (see Leemans, Mon. II 228,1).

82. Champollion (1842), p. 13; Lepsius (1849), pp. 148, 213.
83. Brugsch (1856).
84. Brugsch (1864), pp. 34–52.
85. Brugsch (1883), vol. 2, pp. 392–393.
86. Brugsch (1883), vol. 2, p. 392 (cf. p. 223).
87. Brugsch ((1883), vol. 2, p. 392) also adduces a unique spelling, dating to Dynasty 18, of *3ḫt* as [hieroglyphs], with water-determinative. Its significance is unclear.
88. "Dieser Verbindung des Siriusaufganges mit der Nilschwelle entspricht es, daß die erste der 3 Jahrzeiten, in die das aegyptische Siriusjahr nach dem Muster des alten Bauernjahres . . . zerfiel, die 'Überschwemmung' hieß" (Sethe (1919–1920), p. 294).
89. The length of the solar year is about 365.24220 days. Owing to the precession of the equinoxes, the interval between two risings of the same star is slightly longer. A star year is about 365.25636 days long. But Sirius exhibits a proper motion that makes its star year shorter than that of other stars. In fact, the Sirius year is close to 365¼ days long in antiquity. It has grown longer in more recent centuries. The Sirius year is thus longer than the solar year of about 365.24220 days and shorter than the average star year of about 365.25636 days.
90. von Beckerath (1997), p. 179.
91. Borchardt (1917), p. 3 and Figure 1.
92. See Sethe (1932–1933), p. 2; cf. Sethe (1919–1920), p. 303 with note 1.
93. Neugebauer (1938).
94. Parker (1950), p. 53, § 265.

References

Borchardt, L., 1917, *Die Annalen und die zeitliche Festlegung des Alten Reiches der ägyptischen Geschichte*, Quellen und Forschungen zur Zeitbestimmung der ägyptischen Geschichte 1 (Berlin: Behrend & Co.).

Brinkman, J., 1976, *Materials and Studies for Kassite History*, I: *A Catalogue of Cuneiform Sources Pertaining to Specific Monarchs of the Kassite Dynasty* (Chicago: The Oriental Institute of the University of Chicago).

Brugsch, H., 1856, *Nouvelles recherches sur la division de l'année des anciens Égyptiens, suivies d'un mémoire sur des observations planétaires consignées dans quatre tablettes égyptiennes en écriture démotique* (Berlin: F. Schneider).

——, 1864, *Matériaux pour servir à la reconstruction du calendrier des anciens égyptiens. Partie théorique, accompagnée de treize planches lithographiées* (Leipzig: Librairie J.C. Hinrichs).

——, 1883, *Thesaurus Inscriptionum Aegyptiacarum: Altaegyptische Inschriften.* I: *Astronomische und astrologische Inschriften altaegyptischer Denkmaeler.* II: *Kalendarische Inschriften altaegyptischer Denkmaeler.* (Leipzig: J.C. Hinrichs'sche Buchhandlung). [Reprinted by Akademische Druck- und Verlagsanstalt of Graz in 1968.]

Bunsen, C. C. J., 1845–1857, *Aegyptens Stelle in der Weltgeschichte. Geschichtliche Untersuchung in fünf Büchern.* I: *Weg und Ziel.* II: *Das alte Reich.* III: *Das mittlere und neue Reich.* IV and V: No subtitles (Hamburg: Friedrich Perthes (I: 1845; II: 1844; III: 1845) and Gotha: Friedrich Andreas Perthes (IV: 1856; V: 1857)).

Champollion, J. F. (Champollion le Jeune), 1842, "Mémoire sur les signes employés par les anciens égyptiens à la notation des divisions du temps, dans leurs trois systèmes d'écriture", *Mémoires de l'Institut royal de France, Académie des inscriptions et belles-lettres* 15, 73–134. Read at the Académie on 18 March 1831. The text could not be located for several years (see the note by "J.J.C.F." [Jean Joseph Champollion Figeac, Champollion's elder brother] on pp. 134–135).

Cowley, A. E., 1923, *Aramaic Papyri of the 5th Century B.C.* (Oxford: Clarendon Press).

Crusius, Paulus, 1578, *Liber de epochis seu aeris temporum et imperiorum omnium facultatum studiosis utilissimus* (Basileae: Per Sebastianum Henricpetri).

Depuydt, L., 1995a, "On the Consistency of the Wandering Year as Backbone of Egyptian Chronology", *Journal of the American Research Center in Egypt* 32, 43–58.

——, 1995b, "Regnal Years and Civil Calendar in Achaemenid Egypt", *Journal of Egyptian Archaeology* 81, 151–173.

——, 1995c, "'More Valuable Than All Gold': Ptolemy's Royal Canon and Babylonian Chronology", *Journal of Cuneiform Studies* 47, 97–117.

——, 1997a, *Civil Calendar and Lunar Calendar in Ancient Egypt*, Orientalia Lovaniensia Analecta 77 (Leuven: Uitgeverij Peeters en Departement Oosterse Studies).

——, 1997b, "The Time of Death of Alexander the Great: 11 June 323 B.C. (–322), ca. 4:00–5:00 PM", *Die Welt des Orients* 28, 117–135.

——, 1999, "The Two Problems of the Month Names", *Revue d'Égyptologie* 50, 107–133.

——, 2000, "Sothic Chronology and the Old Kingdom", *Journal of the American Research Center in Egypt* 37, 167–186.

——, 2001, "What Is Certain about the Origin of the Egyptian Civil Calendar?", in H. Győry (ed.), *"Le lotus qui sort de terre": Mélanges offerts à Edith Varga* (Budapest: Museum of Fine Arts), 81–94. [Erratum: At p. 91, line 23, for "is written" read "is written ".]

——, 2002, "The Date of Death of Jesus of Nazareth", *Journal of the American Oriental Society* 122, 466–480.

——, 2005, "The Shifting Foundation of Ancient Chronology", in A.-A. Maravelia (ed.), *Modern Trends in European Egyptology: Papers from a Session Held at the European Association of Archaeologists Ninth Annual Meeting in St. Petersburg 2003,* British Archaeological Reports S1448 (Oxford: Archaeopress), 53–62.

——, 2006a, "Saite and Persian Egypt, 664–322 BC (Dynasties 26–31, Psammetichus I to Alexander's Conquest of Egypt)", in E. Hornung, R. Krauss and D. A. Warburton (eds.), *Ancient Egyptian Chronology*, Handbook of Oriental Studies 833 (Leiden, Boston: Brill), 265–283. [Errata (references are to pages and *lines*): (265,*5*) for "can dates" read "can be dated." (265,*11*) move note 1 from "... detail." to after "... time." in *6*.]

——, 2006b, "Foundations of Day-exact Chronology", in E. Hornung, R. Krauss and D. A. Warburton (eds.), *Ancient Egyptian Chronology*, Handbook of Oriental Studies 833 (Leiden, Boston: Brill), 458–470. [Errata (in addittion to those mentioned in note 41): (466,*15*) for "five days less" read "five days more"; (470,*10*) for "Year 5" read "Year 15".]

Goddio-Von Bomhard, A.-S., 2005, "Sothic Dates in Egyptian Chronology", in A.-A. Maravelia (ed.), *Modern Trends in European Egyptology: Papers from a Session Held at the European Association of Archaeologists Ninth Annual Meeting in St. Petersburg 2003,* British Archaeological Reports S1448 (Oxford: Archaeopress), 63–70.

Goldstine, H. H., 1973, *New and Full Moons 1001 B.C. to A.D. 1651*, Memoirs of the American Philosophical Society Held at Philadelphia for Promoting Useful Knowledge 94 (Philadelphia: American Philosophical Society).

Grafton, A., 1993, *Joseph Scaliger: A Study in the History of Classical Scholarship*, II: *Historical Chronology* (Oxford: Clarendon Press).

Grayson, A. K., 1975, *Assyrian and Babylonian Chronicles* (Locust Valley, N.Y.: J. J. Augustin).

Hagens, G., 1995, "The Assyrian King List and Chronology: A Critique", *Orientalia* 74, 23–41.

Hallo, W. H. and Simpson, W. K., 1998, *The Ancient Near East: A History* (Fort Worth: Harcourt Brace College Publishers, Second edition).

Hess, R. S., 1993, *Amarna Personal Names*, American Schools of Oriental Research Dissertation Series 9 (Winona Lake, Indiana: Eisenbrauns).

Horn, S. L. and Wood, L. H., 1954, "The Fifth-century Jewish Calendar at Elephantine", *Journal of Near Eastern Studies* 13, 3–20.

Ideler, Ludwig, 1806, *Historische Untersuchungen über die astronomischen Beobachtungen der Alten* (Berlin: C. Quien).

——, 1825–1826, *Handbuch der mathematischen und technischen Chronologie, aus den Quellen bearbeitet.* 2 vols. (Berlin: August Rücker).

James, P., 1991/1993, *Centuries of Darkness* (New Brunswick, N.J.: Rutgers University Press, 1993. First published in the United Kingdom in 1991 by Jonathan Cape).

Jonsson, C. O., 1998, *The Gentile Times Reconsidered: Chronology and Christ's Return.* (Atlanta: Commentary Press, Third Edition, Revised and Expanded). [It has been reported to me that there is a fourth edition. I have not seen it.]

Kraeling, E. G. H., 1953, *The Brooklyn Museum Aramaic Papyri: New Documents of the Fifth Century B.C. from the Jewish Colony at Elephantine* (New Haven: Published for the Brooklyn Museum by the Yale University Press).

Kunitzsch, P., 1974, *Der Almagest: Die Syntaxis Mathematica des Claudius Ptolemäus in arabisch-lateinischer Überlieferung* (Wiesbaden: Otto Harrassowitz).

Lepsius, R., 1849, *Die Chronologie der Aegypter. Einleitung und erster Teil: Kritik der Quellen* (Berlin: Nicolaische Buchhandlung).

Luft, U., 1992, *Die chronologische Fixierung des ägyptischen Mittleren Reiches nach dem Tempelarchiv von Illahun*, Österreichische Akademie der Wissenschaften, Philosophisch-historische Klasse, Sitzungsberichte 598; Veröffentlichungen der ägyptischen Kommission 2 (Vienna: Verlag der Österreichischen Akademie der Wissenschaften).

Millard, A., 1994, *The Eponyms of the Assyrian Empire, 910–612 B.C.* (Helsinki: Neo-Assyrian Text Corpus Project).

Moran, W. L., 1992, *The Amarna Letters* (Baltimore and London: The Johns Hopkins University Press).

Murnane, W. J., 1995, *Texts from the Amarna Period in Egypt*, Writings from the Ancient World 5, Edited by E. S. Meltzer (Atlanta: Scholars Press).

Neugebauer, O., 1938, "Die Bedeutungslosigkeit der 'Sothisperiode' für die älteste ägyptische Chronologie", *Acta Orientalia* 17, 169–195.

——, 1955, *Astronomical Cuneiform Texts: Babylonian Ephemerides of the Seleucid Period for the Motion of the Sun, the Moon, and the Planets* (London: Published for the Institute of Advanced Study, Princeton, N.J. by Lund Humphries).

——, 1962, *The Exact Sciences in Antiquity* (New York: Harper & Brothers, second edition). [The first edition of 1951 was published by Princeton University Press, the second of 1957 by Brown University Press.]

——, 1975, *A History of Mathematical Astronomy*, Studies in the History of Mathematics and Physical Sciences 1 (New York, Heidelberg, Berlin: Springer-Verlag).

Parker, R. A., 1950, *The Calendars of Ancient Egypt*, Studies in Ancient Oriental Civilization 26 (Chicago: The University of Chicago Press).

——, and Dubberstein, W. H., 1956, *Babylonian Chronology 626 B.C. – A.D. 75*, Brown University Studies 19 (Providence: Brown University Press).

Poebel, A., 1942/1943, "The Assyrian King List from Khorsabad", *Journal of Near Eastern Studies* 1, 247–306, 460–492 and 2, 56–90.

Porten, B., 1990, "The Calendar of Aramaic Texts from Achaemenid and Ptolemaic Egypt", in S. Shaked and A. Netzer (eds.), *Irano-Judaica*, II: *Studies Relating to Jewish Contacts with Persian Culture throughout the Ages* (Jerusalem: Makhon Ben-Tsevi), 13–32.

——, and Yardeni, A., 1986–1993, *Textbook of Aramaic Documents from Ancient Egypt.* First 3 vols. of 4 (vols. 2 and 3 accompanied by folders with plates) (Jerusalem: Akademon. Distributed by Eisenbrauns: Winona Lake, Indiana).

Postgate, N., 1991, "The Chronology of Assyria – An Insurmountable Obstacle", *Cambridge Archaeological Journal* 1, 244–246.

Rühl, F., 1897, *Chronologie des Mittelalters und der Neuzeit* (Berlin: Verlag von Reuther & Reichard).

Saliba, G., 1994, "The Development of Astronomy in Medieval Islamic Society", in *A History of Arabic Astronomy: Planetary Theories during the Golden Age of Islam* (New York and London: New York University Press), 51–65. [This article first appeared in *Arab Studies Quarterly* 4 (1982), 211–215.]

Salvolini, F., 1833, *Des principales expressions qui servent à la notation des dates sur les monuments de l'ancienne Égypte* (Paris: Librairie Orientale de Dondey-Dupré).

Scaliger, Joseph Justus, 1583, *Opus novum de emendatione temporum in octo libros tributum* (Paris (Lutetiae): Apud Mamertum Patissonium Typographum Regis, In Officina Roberti Stephani). [The edition of 1593

(Francofurti: Apud Ioannem Wechelum, Sumtibus Nicolai Bassaei Typographi) is a reprint of this first edition of 1583.]

——, 1598, *Opus de emendatione temporum.* Second ("new") edition, augmented and revised (*castigatius et multis partibus auctius ut novum videri possit*) (Leiden (Lugduni Batavorum): Ex Officina Plantiniana Francisci Raphelengij).

——, 1629, *Opus de emendatione temporum.* Third ("last") edition, augmented and revised (Geneva (Coloniae Allobrogorum): Typis Roverianis).

Schmidtke, F., 1952, *Der Aufbau der babylonischen Chronologie* (Münster: Aschendorffsche Verlagsbuchhandlung).

Sethe, K., 1900, "Der Name der Überschwemmungszeit", *Zeitschrift für ägyptische Sprache und Altertumskunde* 38, 103–106.

——, 1904, "Die endgültige Lesung für den Namen der Überschwemmungszeit", *Zeitschrift für ägyptische Sprache und Altertumskunde* 41, 89–90.

——, 1919–1920, *Die Zeitrechnung der alten Aegypter im Verhältnis zu der der andern Völker*, Nachrichten von der Königlichen Gesellschaft der Wissenschaften zu Göttingen, Philologisch-historische Klasse, 1919.3: 287–320; 1920.1: 28–55; 1920.2: 97–141 (Berlin: Weidmannsche Buchhandlung).

——, 1932–1933, *Urkunden des Alten Reichs.* 4 fascicles (fascicles 1–2 in second edition), Urkunden des ägyptischen Altertums 1 (Leipzig: J.C. Hinrichs).

Swerdlow, N. M. and Neugebauer, O., 1984, *Mathematical Astronomy in Copernicus's De Revolutionibus*, Studies in the History of Mathematics and Physical Sciences 10 (New York, Berlin, Heidelberg, and Tokyo: Springer-Verlag).

Tihon, A., 1994, "L'astronomie byzantine (du Ve au XVe siècle)", in *Études d'astronomie byzantine*, Variorum Collected Studies Series CS 454 (Aldershot, Hampshire: Variorum), Article no. 1. [This article first appeared in *Byzantion: Revue internationale des études byzantines* 51 (1981), 603–624.]

von Beckerath, J., 1994, *Chronologie des ägyptischen Neuen Reiches*, Hildesheimer ägyptologische Beiträge 39 (Hildesheim: Gerstenberg).

——, 1997, *Chronologie des Pharaonischen Ägypten*, Münchner ägyptologische Studien 46 (Mainz: Philipp von Zabern).

von Gutschmid, A., 1889, in F. Rühl (ed.), *Kleine Schriften.* I: *Schriften zur Aegyptologie und zur Geschichte der griechischen Chronographie* (Leipzig: B.G. Teubner), 365–371. [Contains a review of Heinrich Brugsch's *Histoire d'Égypte dès les premiers temps de son existence jusqu'à nos jours*, I: *L'Égypte sous les rois indigènes* (Leipzig: J.C. Hinrichs'sche Buchhandlung, 1859). The review was first published in the *Literarisches Centralblatt* of 1861, at pp. 51–54.]

Walker, C. B. F., 1997, "Achaemenid Chronology and the Babylonian Sources", in J.Curtis (ed.), *Mesopotamia and Iran in the Persian Period: Conquest and Imperialism* (London: British Museum Press), 17–25.

Weidner, E. F., 1944, "Die Königsliste aus Khorsābād", *Archiv für Orientforschung* 14, 362–369.

Wells, R. A., 2001, "The Role of Astronomical Techniques in Ancient Egyptian Chronology: The Use of Lunar Month Lengths in Absolute Dating", in J. M. Steele and A. Imhausen (eds.), *Under One Sky: Astronomy and Mathematics in the Ancient Near East*, Alter Orient und Altes Testament 297 (Münster: Ugarit-Verlag), 460–472.

Young, R. C., 2004, "When did Jerusalem Fall?", *Journal of the Evangelical Theological Society* 47, 21–38.

The 360-Day Year in Mesopotamia

Lis Brack-Bernsen

The Cultic or Civil Calendar and the "Administrative Calendar"

From ancient Mesopotamia we know two different "calendars": the cultic or civil calendar and the "administrative calendar".[1] In both calendars, the year was the largest unit. In the cultic calendar, the year was divided into lunar months of 29 or 30 days. In the administrative calendar each month was put equal to 30 days. As we shall see, the administrative calendar was only used in the context of making calculations and never for dating. This latter use was taken care of by the cultic calendar which was defined by the solar year and by lunar phases: the first day of a new month began on the evening when the new crescent became visible for the first time after conjunction. First visibility takes place either 29 or 30 days after the previous first visibility. Since the sequence of 29- and 30-day months is highly irregular the cultic calendar was troublesome to handle. The mean value of the synodic month (the time between two consecutive conjunctions or oppositions) is approximately 29.53 days, so that 12 lunar months equals about 354.37 days, some 10.88 days shorter than the solar year. Therefore, on the average, an intercalated month was necessary every three years or so in order to keep the months in tune with the seasons.

This paper shall present textual evidence for the claim that both systems of time measurements were used simultaneously from the early dynastic time (*ca.* 2600 BC) until around 1000 BC when the astronomical-astrological compendium MUL.APIN was composed. Prior to the early dynastic period there is insufficient evidence to say for certain whether the lunar calendar was used; the protoliterate texts (from *ca.* 3200 BC) have not been understood sufficiently—one can only speculate, that some signs in Uruk III texts may refer to the kind of cultic lunar month known from later periods. However, the number systems in these very early texts are sufficiently well understood to enable us to understand the calculations they contain and to establish that rations in bureaucratic texts were calculated of the basis of the administrative year. The paper shall also clarify how (i.e., for what or in which way) the two calendars were used: the cultic calendar for dating and the administrative one for centralized bookkeeping and later also for astronomy. For example, in MUL.APIN, a canonical astronomical/astrological text perhaps dating from around 1000 BC, we find astronomical schemes in which the ideal year of 360 days is used, together with indications of how to adjust schematic values according to observations.

I start with Ur III texts, which are understood profoundly, and then go backwards in time to the earlier texts. The knowledge gained from the Ur III texts makes it is easier to understand and explain older texts. For textual evidence from earliest times until the end of Ur III times (2000 BC), I rely mostly on R. K. Englund's investigations in "Administrative

Timekeeping in Ancient Mesopotamia" (1988), a careful and systematic analysis of all available texts of interest together with many useful references. The conclusions: Bureaucratic texts from the Ur III period show unambiguously that and how the artificial administrative year was used—namely as a fixed grid opposed upon the varying lunar calendar. In bureaucratic calculations the duration of each month (independent of its actual length) was simply always taken to be 30 days. The same accounting practice is documented for pre-Sargonic Girsu (2400 BC) by feed texts (i.e., texts listing the daily and monthly amount of grain necessary for feeding different types of animals). The tradition can, however, be followed further back. Texts from the protoliterate period use a system of time notation which is very similar to that of the texts from the pre-Sargonic period. And archaic fodder and ration texts firmly establish the correctness of archaic time notation as proposed by Vaiman in the early 1970s.[2] Englund has thus been able to conclude "that in the protoliterate period the same system of administrative time recording was employed as was the notational basis 1000 years later". This means that also at these early times, a month was always calculated as having 30 days and a year as 360 days in administrative records.

There is, hence, clear textual evidence for the administrative calendar from protoliterate time and onwards. This is not the case for the cultic calendar. The number symbols and calculations on protoliterate texts have been understood; but not the other signs which may refer to cultic events connected to the year or lunar phases. The understanding of archaic texts is too limited. But still, most scholars (including myself) tend to believe that daily life from the earliest beginning, i.e., from proto-literate time, was regulated by the sun and by lunar phases, while at the same time the artificial time measuring system was used in centralized bookkeeping. One argument for this claim is the continuity of cult and tradition in Mesopotamia, and the cultic year is evident from the earliest tablets (from 2600 BC), where also the text can be deciphered, and onwards. Another argument is that the administrative calendar is an approximation to the lunisolar calendar. But we have no textual evidence for this conviction. Another possibility would be that the administrative year was used as a calendar in archaic times. This would mean that each month had 30 days, so that the lunar phases were not in tune with the calendar month. New- and full-moon could take place on every day of the month. It is possible, but to me not very plausible, that in archaic times a schematic month and not direct observations, was the basis of time regulation. At least from 2600 BC onwards, we know that cultic events took place on days with special lunar phases and that the civil astronomical lunisolar calendar was used for timekeeping.

In *Der kultische Kalender der Ur III-Zeit*, Sallaberger (1993) analyses Ur III-texts concerning cyclical cult festivals. He comes to the same conclusion that the cultic calendar regulated life while the 360 day year was used for accounting. On p.11 he writes:

> Ein Mondmonat (synodischer Monat) dauert 29d 12h 44' 2,9", es wären also etwa abwechselnd Monate zu 29 oder 30 Tagen zu erwarten. In Ur III-Urkunden über Arbeiter, Rationen, Futter oder in Bilanzrechnungen wird dagegen ein normierter 30-Tage-Monate verwendet. Dieser 30-tägige Normmonat dient offenkundig nur der Berechnung im administrativen Bereich und gibt keine Auskunft über den allgemein gebrauchten "bürgerlichen" Kalender.

I refer to the work of Eleanor Robson (1999) and of Jöran Friberg *et al.* (1987–90) in order to demonstrate the continuous tradition of the artificial administrative year (of 12 × 30 days) until it appears in astronomical tables in the astronomical/astrological compendium

MUL.APIN. Eleanor Robson has shown that the training of Old Babylonian scribes demonstrates a continuous tradition leading back to Ur III administrative practices and mathematics: work rates known from Ur III times show up in Old Babylonian school mathematics and coefficient lists. And Jöran Friberg has been able to document that an unbroken tradition ran from the well-documented mathematics of the Old Babylonian period to the mathematics of the Late Babylonian period more than a millennium later.

Since the argumentation on the archaic texts depends heavily on the archaic systems of number notation, I shall start with a short outline of early counting.

Early Accounting Techniques and Writing

Tokens (and counting symbols) have been used since the 8th millennium BC. D. Schmandt-Besserat (1977) noted that many decorated tokens from the 4th millennium bear a striking resemblance to signs on the earliest tablets, so she interpreted them as three-dimensional precursors of the two-dimensional proto-cuneiform signs. Whilst some of her conclusions remain controversial, what has been generally accepted is her argument that simple tokens served as precursors of number signs.

Writing was invented toward the end of the fourth millennium BC within a growing society with an urgent need of economic administration: the purpose of the earliest clay tablets with proto-writing was to record numerical information.

In archaic administrative texts one finds a variety of number symbols and counting systems. The number symbols were produced by pressing round sticks of different diameters perpendicular or slanted into clay (see figure 1). A. Vaiman and J. Friberg were the first to investigate the archaic accounting texts.[3] Friberg showed that the number symbols were dependent on the context: the •, when used for counting items or animals, meant 10 while it was used for the number 6 within the recording of quantities of grain. Later a group in Berlin, analyzing a large amount of archaic Texts from Uruk, was able to identify 60 different number signs (N_1, N_2, ... N_{60}), which were used in five basic and a further five derived number sign systems. The symbols show a great variation as if they were slightly altered for different purposes. Analysis of texts of archaic bookkeeping, on which the balance, i.e., the sum of all contributions is recorded on the back, led to the recognition of all the different counting systems.[4]

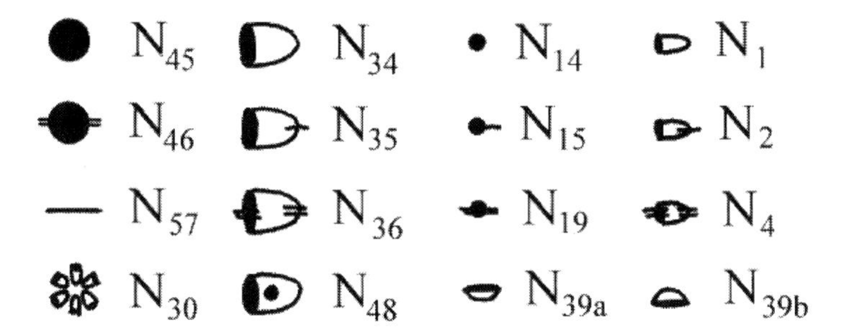

FIGURE 1. Some of the sixty different number symbols found on archaic tablets.

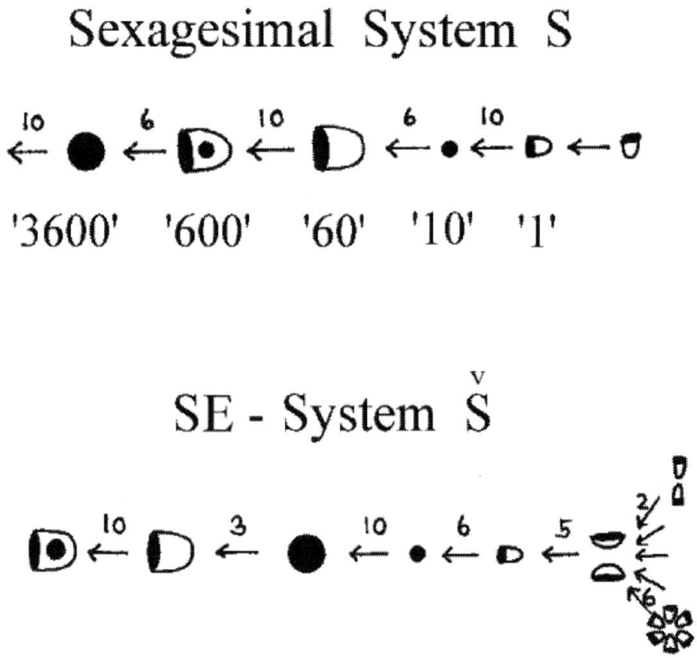

Sexagesimal System S

10 6 10 6 10

'3600' '600' '60' '10' '1'

SE - System Š

10 3 10 6 5 2 6

FIGURE 2. Two of the counting systems: the sexagesimal system S (used for counting discrere objects) and the system Š (used for the recording of volumes of grains, mostly barley).

The order in which the number symbols occurred on the tablets helped identify higher counting units, and since the sum of all contributions was recorded, the factor between the different number symbols could be determined. At this place we only show two of the archaic numbersign systems: the sexagesimal system S, which was used for counting discrete objects, and the ŠE-system Š, which was used for recording volumes of grains, mostly barley (see figure 2). Note the different interpretation of the symbol • = N_{14}: In system S, • equals 10 units (• = 10 N_1) while in system Š it must be read as 6 basic units (• = 6 N_1). We see that number symbols were dependent on the context. This has been discovered only rather recently by Friberg (1978).

The archaic system S was later developed into the well-known sexagesimal place value system used in all cuneiform texts from Old Babylonian times onwards. Here all numbers from 1 to 59 were written by means of the two signs 𒁹 and 𒌋. The cone, 𒁹, which is the later version of N_1 for one, and the angle, 𒌋, which is the later version of N_{14} = • for 10. For instance, the number twenty-three would be written as 𒎙𒐈. That all numbers could be written by means of only the sign for 1 and that for 10, led earlier scholars to the incorrect conclusion that at early times, a decimal and a sexagesimal number system have coexisted. However, the postulate of two competing systems, the sexagesimal and the decimal system, originates in the incorrect reading of • = 10. We now know that • can be read as 6 in case of measuring grain, and as 10 when used for counting things, and also as 18 when • was used for notations on the sizes of fields.

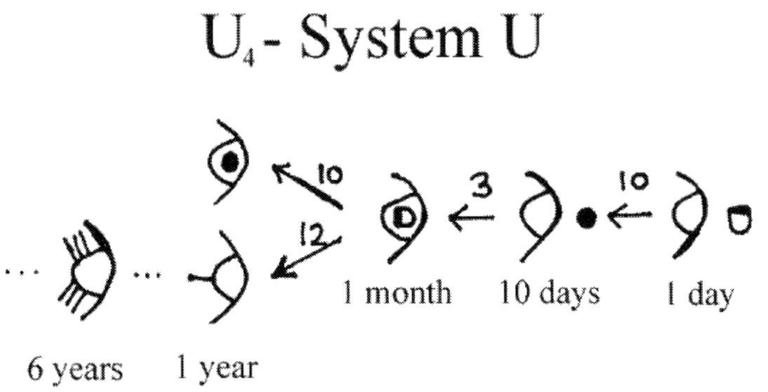

U$_4$- System U

10 3 10

12

1 month 10 days 1 day

6 years 1 year

FIGURE 3. System U was used for recording time intervals. The common sign U$_4$ is assumed to represent the sun rising among the eastern mountains of Mesopotamia. Note that the sexagesimal system S was used for counting months and days. Vaiman proposed that numbers inside U$_4$ counted months and numbers after the sign U$_4$ counted days, while strokes in front of the sign U$_4$ counted years.

Archaic administrative texts exhibit a variety of counting systems. Throughout the third millennium BC, concurrently with the increasing demands of large-scale state bureaucracies, the counting and measuring systems were gradually revised and the way of writing numbers was changed. At the end the sexagesimal place value system, as we know it from astronomical and mathematical texts, was derived. In the early Ur III period, the king Šulgi introduced major bureaucratic reforms, the sexagesimal system sometimes being used in calculations although the results were given in the traditional metrological system. At the same time mathematical methods for accounting were developed and standardized, for example for managing labour and resources, or for the calculation of expected (or needed) labour, but the calculations were still based on the administrative year of 12 months of 30 days. Scribes were also trained in mathematical methods, enabling them to fulfil duties in accounting and administrating. From the earliest phase and throughout the history of cuneiform, tablets were produced for the education of scribes. Many such tables happen to survive from the Old Babylonian period, the Old Babylonian mathematical texts being a rich source of early mathematics. Mathematics was an integral part of Mesopotamian scribal culture.[5]

Next I shall present the textual evidence for the claim that the administrative and civil calendar were used simultaneously.

Time Notations in Ur III Texts

The administrative calendar and the cultic or civil calendar are both clearly attested in administrative texts from Ur III. More than that, many texts from Ur III times (*ca.* 2100 BC) exhibit simultaneously the use of these two parallel systems of time division. But as we shall see, daily life was regulated by the lunar calendar while the administrative calendar was used only for bookkeeping.

During the Ur III period cultic festivals are attested at significant times during the agricultural year and on notable days in the cycle of lunar phases. As a result the cultic year was naturally divided into 12 (or 13) synodic months, where the beginning of a new month could be found by observing the moon. The astronomical lunisolar calendar at this period was quite irregular as no easy system for organising the intercalation of the 13th month had been found. Furthermore, the duration of the synodic month, which can be either 29 or 30 days, varies in a very irregular pattern. Nevertheless, life and cultic events were structured by this cultic calendar.

At the same time, scribal computations were carried out by an administrative system of time division in which the year consisted of 12 months of 30 days each. In contrast to the irregular lunisolar calendar, the 360-day year is very practical for bureaucratic calculations—it simplifies calculations and at the same time it increases the state's demands on labour. Independent of its actual length, each synodic month was simply calculated as having 30 days. As an example of this practice we can consider an Ur III text published by Englund (1988), p.126. The text calculates the number of workdays which a group of 36 female workers should deliver during a year of 12 (lunar) months.

> 36 female workers
> from the "harvest(-festival)" (month 1, Umma calendar)
> through "Tammuz(-festival)" (month12),
> performance involved: 12,960 days

The total number of workdays, 12,960 days, equals $36 \times 12 \times 30$. We see that the expected labour was calculated as the product of the number of workers and the period recorded. The text also illustrates that the accounting year (of 12 months from month 1 through to month 12) was calculated as having 360 days.

By similar texts, Englund (1988), p.128 has shown, that a diri year (a year of 13 months) was calculated as having 390 days. For example:

> 20 ^{gis}gid$_2$.da–workers,
> ...,
> and 4 porters
> from "Harvest(-festival)" (month 11, Girsu calendar)
> of the year "The throne of Enlil was constructed" (Amar-Zu'en 3)
> through "Amar.a.a.si" (month 10)
> of the year "Enmahgalana was installed as priestess of Nanna" (Amar-Zu'en 4)
> performance involved: 9360 workdays, male workers,
> it is (a period of) 13 months,
> including one extra month.

Again the text covers the period of a year, staring with the beginning of month 11 and running until the end of month 10 the following year. But the text indicates (by "13 months") that in this case, an extra month has been inserted. Since $24 \times 30 \times 13 = 9360$, we can conclude that this civil diri year was reckoned as having $30 \times 13 = 390$ days.

Note the month name "Harvest(-festival)" of the civil Girsu calendar (agricultural festivals were celebrated by sowing and harvest). The civil/agricultural calendar was evidently used for recording dates on this administrative tablet. Englund provides other examples of bookkeeping tablets where the civil/agricultural calendar and the parallel administrative calendar were used simultaneously within the same text.

The division of the year into 12 lunar months in the cultic or civil calendar was dictated by the moon cult: Ur III texts tell that cultic offerings took place on the 7th, 15th and new moon days. Evidently, the moon phases first quarter, full moon and new moon are meant here. It is highly improbable that the new moon celebration should not take place at real new moon but, instead, on the first day of an administrative 30-day month. Had the 360-day year been utilized as a calendar, then the first day of the administrative month could fall on every lunar phase, since the 30 day month shifts 0.5 day per month in comparison to the lunar month. In addition, Ur III deliveries of offerings for the 15th (full moon) are received at the end of the 12th day, while new moon offerings are recorded only on the 27th to the 30th days of particular months. Here, evidently, the civil calendar is used: within the administrative month of 30 days, the full and new moon would wander and could take place at every date throughout the month.[6]

In his dissertation from 1993, Sallaberger analyses Ur III archives in the search for evidence for the civil lunisolar calendar. In texts connected to agricultural and lunar cults or festivals, he was able to find many examples of evidence for use of the lunar/agricultural calendar, i.e., the civil/cultic calendar. Let me just summarize some of the evidence in the following.

Each month in Gula either 29 or 30 carcasses of dead cattle were delivered, a clear hint to the changing length 29 or 30 days of the lunar month. The sum of delivered carcasses over some whole years were recorded to be 384, 354 or 353, respectively. Thirteen lunar months have 384 days, while 12 months encompass 354 days. Evidently, the civil calendar was used here and not the administrative calendar. Finally, texts on cultic actions testify that the month was regulated by the real moon: some purifying bath rites were performed by the queen on the "day at which the moon rests" (= "Schwarzer Mondtag" = day of conjunction). Obviously, these cultic actions must have been performed at days where the moon was absent. Lists shows that carcasses of small cattle for these rites were always delivered during the days from the 24th to the 27th of the month, indicating that the rites took place at the end of the lunar month.

Further evidence for the fact that the cultic year and not the administrative year was used for regulating civil life can be found by the frequency of "diri" years with a 13th intercalated month. The civil calendar requires an intercalation every three years, while the administrative year of 360 days would only require an additional month every six years. A text from the time of Šulgi writes that six intercalations had taken place within 16 years which is only possible in the cultic calendar:

> Tablet basket:
> accounts
> of Nalu
> from the "Gazelle-eating (festival)" (month1, Drehem calendar)
> of the year "Simurum was destroyed for the second time" (= Šulgi 26)
> through the extra month "Harvest(-festival)" (month 13)
> of the year following the year following "The house/temple of Drehem was built" (= Šulgi 41),
> It is (a period of) 16 years, including 6 extra months

This and other Ur III texts confirm the pattern of an intercalation (on the average) every three years, indicating that the cultic calendar was used.[7] We see that daily life and cultic events were organized by the synodic calendar, while at the same time bureaucratic calculations utilized the administrative calendar. In their dependence on the parallel lunisolar year, Ur III administrators inserted into the administrative year an intercalary month of 30 days, again in principle, every three years, in tune with the intercalations for the cultic year. In other words:

each lunar month was just reckoned as having 30 days independently of its actual duration.

Finally, the administrative bakery text TUT 102 (originating from Telloh and dated to Ur III times) can be mentioned.[8] It records 59 (60 lal 1) days in months 3–4 of the Lagash calendar. Obviously, in this administrative text the cultic calendar is attested.

To conclude: Ur III texts provide clear evidence for the simultaneous use of the two calendars. The administrative calendar facilitated calculations and increased the state's demands. I would describe the artificial year (with months of 30 days) as a fixed grid opposed to the real and variable year and adjusted to it by adding an extra month in tune with the intercalations necessary for the civil year. For dating and normal/cultic life, the lunisolar calendar alone was used. The 360-day year was only used in connection with accounting.

Presargonic Texts from Girsu

The use of the 360 day year can be demonstrated also for the presargonic period (*ca.* 2500 BC – 2350 BC) through ration and feed texts, which calculate the food needed for fattening swine of different ages. Knowing the pre-Sargonic Girsu gur system, Englund (1988) pp. 141–142 could show how, for example, the monthly amount grain needed to feed the largest wild boar was found by multiplying the daily need by 30. The animals are fed according to their age which is given in years (e.g., šah$_2$ .. mu.2 "swine in its second year"). The older the animal, the more food it was given. The barley needed for feeding N animals during a month is found by multiplying the daily ration R_i (for age$_i$ swine) by N times 30:

$$\text{Monthly need for N (ration } R_i\text{)-swine} = N \times 30 \times R_i$$

The yearly need equals 12 × 30 × the daily need. From texts such as this we can conclude that the year of 12 × 30 days is well attested as an administrative time unit in pre-Sargonic Girsu. In other words, in texts calculating the food for animals, the administrative calendar was used.

However, we also have evidence for the use of the lunisolar calendar in texts from pre-Sargonic Girsu. Many of the 24th century texts, recording quantities of feed for animals, show a mixture of the two calendar systems within the text.[9] For example, the colophon of an administrative text mentions the "Malt-eating festival" month of the cultic calendar. This month is administratively reckoned as the ninth 30-day rationing period. Obviously, the "malt festival" month of the cultic calendar has here been identified with the ninth schematic month. Other presargonic texts show that the artificial administrative and the traditional agricultural year have existed side by side, and that the administrative year sometimes would have an extra 13th month of 30 days. As it was the case for the Ur III period, it is also true for the presargonic period that the normal year of 12 month was reckoned as 12 × 30 days = 360 days, while the intercalary years would have 13 months of 30 days which equals 390 days.

The pre-Sargonic texts are important for the understanding of archaic texts. The cuneiform signs for one month and one day (see figure 4), used in ration and feed texts from pre-Sargonic Girsu, are so similar to those proposed by Vaiman for the Uruk IV and III periods, that his reading is clearly established.

Archaic Protoliterate Texts

How was the situation in even earlier times? Archaic protoliterate texts may contain information. However, from such tables only the numbers and time notations can be read with certainty,

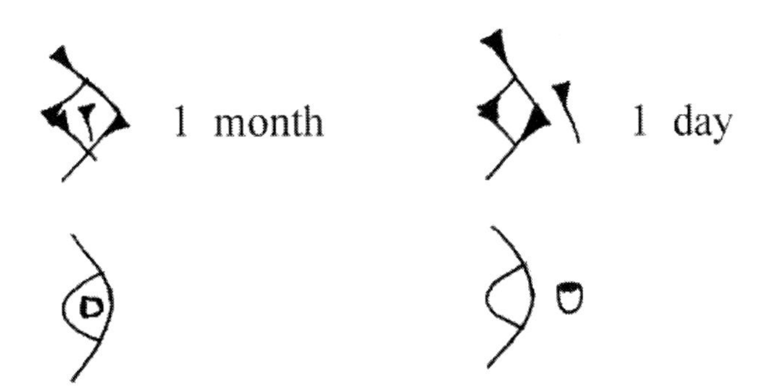

FIGURE 4. The signs (top line) for 1 month and for 1 day used in feed and ration texts from pre-Sargonic Girsu. These signs are very similar to the archaic time symbols (shown underneath), which Vaiman proposed to mean 1 month and 1 day respectively.

while it is much more difficult to understand the text. Archaic texts from Uruk (3200 BC – 3000 BC) concern centralized bookkeeping. They document the same artificial division of the year into 12 months of 30 days each that we know from later periods. Analyses of archaic accounting texts have led to the conclusion that the same system of administrative time reckoning was employed to that used 1000 years later. Exactly as we know it from later times, the archaic ration or fodder texts are calculated on the basis of the 30-day month. The rations for, say, M months are always calculated as M × 30 times the daily ration. The key for understanding the archaic administrative texts is the knowledge of archaic number-sign systems and archaic time notation. This knowledge allows us to find the factor between rations for different time intervals and to establish the great similarity with accounting methods found in later texts (see figure 5).

Note that, according to the number system Š (see fig. 2), 4 N_{14} equals 4 × 6N_1 (= 24 monthly rations), which again equals 24 × 30 N_{30} = 720 N_{30} = 720 daily rations. Many texts make it apparent that the NINDA was the daily grain ration for a worker. It represented a grain quantity equal to the sign $N_{30} = \frac{1}{30}N_1$.

A group in Berlin working on archaic administrative texts has proposed that the bevelled-rim bowl with an average capacity of 0.8 litre was used as a model for the pictogram NINDA, which represented one day's grain ration for a worker. They propose to read the head with a NINDA bowl in front of it as the sign for ration.[10]

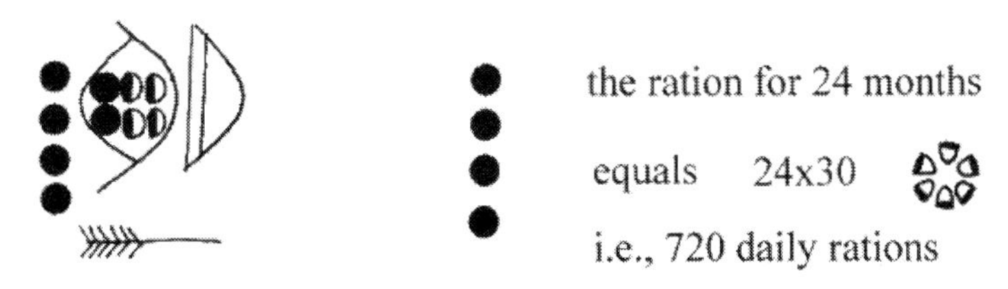

FIGURE 5. Excerpt from an archaic text (shown on the left side). We see the gran quantity of 4 N_{14}, the sign for 24 months and the NINDA bowl. The ration corresponding to 24 months is, indeed, 4 N_{14}: according to the SE-system S (figure 2), 1 N_{14} = 6 N_1, 1 N_1 = 5 N_{39}, and 1 N_{39} = 6 N_{30}. Here the administrative calendar has evidently been used for calculating rations.

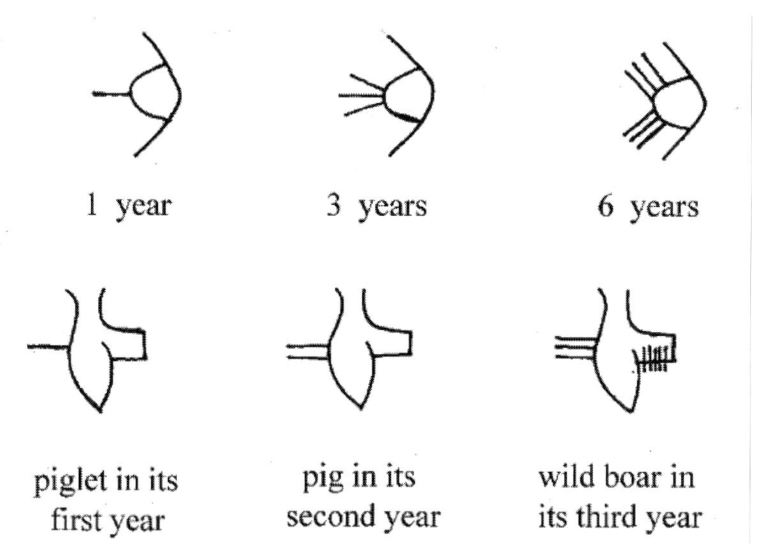

FIGURE 6. Below, the signs for piglets in their first and second years plus wild boar in the third year. The dash in front of the sign for pig is here used for indicating the age. In agreement with Vaiman's proposal, the upper signs can be read as one year, three years and six years.

Due to the still limited understanding of the archaic texts, Englund is still not able to offer solid evidence of the cultic/agricultural calendar being used in the protoliterate period. This would presuppose a higher linguistic level of decipherment than afforded by the now solid clarification of the arithmetical operations behind administrative time keeping in archaic sources. However, both J. Friberg as well as Englund see textual hints to the cultic calendar in archaic times although they are currently unable to give a semantic justification.[11] For archaic texts which might contain evidence of a cultic calendar, see Englund (1998), p. 127 and (2001), pp. 20–21). Based on other texts, Englund presents reasonable arguments for the reading of a colophon as a date, referring to the month "new growth festival".

If we accept the judgment of Englund and Friberg that the archaic texts refer to lunar festival-months, then we must conclude that the lunisolar calendar was the primary time counting system and must have been used first. This would indicate that in archaic times, as in later periods, the 360 days year consisting in 12 months of 30 days was just a practical approximation to the year of 12 synodic months. It seems plausible, since one cannot construct an approximation out of the blue.

The final proof for the correctness of Vaiman's reconstruction of archaic time notation can be found in texts on feeding pigs. As was the case with ration texts from presargonic Girsu, the archaic texts list pigs according to their age in years. Englund has shown that the accounting system, known from Girsu texts, was already used in archaic texts, so that the signs for 1-, 2-, and 3-year old pigs could be identified. The dash, which Vaiman proposed for indicating years in front of the "sun at horizon sign", is here used in front of the sign for pig to indicate its age. Piglets in their first year were listed by the sign for pig with one stroke in front of it, and pigs in their second (or third) year were written by two (or three) strokes in front of the pig sign (see figure 6).

We conclude that the continuous and parallel use of the two calendar has been shown all way through from 2600 BC to 2100 BC, while in the archaic texts, only the administrative calendar is clearly demonstrated. We shall now look for evidence for a continuation of this tradition and show that the scribes sometimes used the approximate year of 12 × 30 days in later periods as well.

The Calendar in the Old Babylonian Period and After

Most scholars assume that the lunisolar calendar was used by the Babylonians all way through until the end of the Seleucid Era. The reason for this is that the real month was basis of the civil calendar in Ur III times and also of the calendar used in the Late Babylonian Astronomical Diaries to record astronomical observations and predictions, historical events, and prices. Whilst there is no direct Old Babylonia evidence for the 29-day month it seems very improbable that during the intermediate period between Ur III and the last few centuries BC a switch would have been made to using the schematic year as the civil calender.

A hint to the use of the lunisolar calender during the Old Babylonian period was found by Landsberger (1949) in the text CT 2, 18, an administrative doument calculating wages and other disbursements. Some ud.da.gid.da days, recorded in this text in connection with time intervals, were interpreted by Landsberger as representing the correction between the schematic and the true (lunisolar) calendar. This would seem to be a nice confirmation of the parallel use of the two calendars. However, it must be noted that the text is corrupt. Michel Tanret (2004) offered a new interpretation of the text CT 2, 18. Referring to inconsistencies in the text, he came to the conclusion that the calendar year had 360 days, so he questions the use of a real lunar calendar in Old Babylonian times. However, Peter Huber (private communication) has given several strong arguments for a lunisolar calendar during the Old Babylonian period: for example, during this time (as well as for Ur III times), an intercalary month was inserted, on the average, every three years, as we would expect in a lunisolar calendar. The schematic calendar of 360 days would only need an intercalation every six years.

The text CT 2, 18 is too corrupt to settle the matter, but Landsberger's interpretation of the ud.da.gid.da days seems to me to be correct. In any case, the evidence for the real month and the lunisolar year to be the basis of the calendar in Ur III and in Old Babylonian times is strong. Therefore, I take it for granted, that the civil calendar as means of dating was used continuously from Ur III through Old Babylonian times to the end of the Seleucid Era. We shall now look for textual evidence for parallel utilization of the approximate year of 12 × 30 days during these later periods.

The Archaic Administrative Calendar and the "Ideal Year"

The civil calendar as means of dating is attested continuously from Old Babylonian times to the end of the Seleucid Era. In the astronomical/astrological compendia *Enūma Anu Enlil* tablet XIV and MUL.APIN, a so-called "ideal" or "schematic" year of 12 × 30 days is used. Omen texts, letters and reports show that it was interpreted as a good omen when a new month started on day 1 (which means that the month before had 30 days). And it was a bad omen when a new month started on day 30 (indicating that the former month had only 29 days). Obviously, it was taken as a good sign when nature behaved as the "ideal" 360-day year and a bad sign when nature deviated from the "ideal calendar".[12] Great efforts were made for finding rules enabling the astronomers to predict the length of a month to come. The Late Babylonian procedure text TU11 gives several methods for the prediction of "full" months

Month	Day	Zodiacal Sign	Degree	Advice: "... you brush"
IV	1	1 = ♈	7	sheep-blood, sheep-fat and sheep-hair you brush
IV	2	10 = ♑	14	goat-blood, goat-fat and goat-hair you brush
IV	3	7 = ♎	21	"empty place"
IV	4	4 = ♋	28	crab-blood and crab-fat you brush
IV	5	2 = ♉	5	bull-blood, bull-fat and bull-hair your brush
.	.	. . .	.	.
.	.	. . .	.	.
IV	30	4 = ♋	30	crab-blood and crab-fat you brush
V	1	2 = ♉	7	bull-blood, bull-fat and bull-hair you brush

TABLE 1. The *Kalendertext* scheme for Month IV: through columns one to four each day of Month IV is associated with a position in the zodiac, while column five mentions blod, fat, hair or other body parts of the animal that corresponds to the zodiacal sign relevant to the day in question.

(i.e., of 30 days) or "hollow" (29 days) months.[13] Many mathematical astronomical tablets of the Seleucid period aim at calculating the day of new crescent and the time NA_N from sunset to the setting of the crescent.[14] In the following I will show that the lunisolar calendar and the "ideal" calendar coexisted all way through from Old Babylonian times (1800 BC) to around 300 BC, and that the "ideal" calendar just is a continuation of the artificial year of accounting.

The so-called *Kalendertexts* may give a good illustration for the coexistence of the two calendars. The different types of *Kalendertext* are explained in Brack-Bernsen and Steele (2004). One type of *Kalendertext*, relates dates to positions in the zodiac. The text, W.22704,[15] from the end of the 4th century BC, may serve as an example of this type. The text has the form of a scheme consisting in 5 columns. In the first column, the name of Month IV (ŠU) is written (or the repetition sign MIN after line 1). Column two records the day numbers from 1 to 30. The entries in columns three and four record a zodiacal signs and a number of degrees within that sign, and finally the last column list for each entry ingredients that are usually linked to the zodiacal sign given in column three. For example, when the sign is Taurus, then blood, fat, and hair of a bull are mentioned (see table 1).[16]

For each day of Month IV the scheme indicates the ingredients which may be used in some ritual actions. But in the *Kalendertext*-schemes, the months are always 30 days long, so we are not dealing here with a real lunar calendar, but instead with the schematic or "ideal calendar" of twelve 30-day months. The *Kalendertext*-schemes are constructed on the basis of the practical "ideal year" and the zodiacal positions relating to dates were given for 30 days of each month. But we know that the calendar used at this time was the well-known astronomical (civil/cultic) lunisolar calendar, with months of 29 or 30 days length. Sometimes, the scheme would have a day more than the real month; but of course, the *Kalendertext*-scheme could still be used to determine a zodiacal position for each day of the real calendar.

As mentioned above, one encounters the so-called "ideal year" in *Enūma Anu Enlil* tablet XIV and MUL.APIN. In idealized schemes, the lengths of day and night are given as functions of the date within the schematic year consisting of 12 months of 30 days each. In these compendia, the length of night is then used as a generating function for finding the "visibility times" for the moon. The Late Babylonian procedure text TU11 testifies to many more astronomical quantities derived from the length of day or night. We see that the ideal year was utilized heavily in the early formation of numerical astronomical theory.[17] Even in the so-called ACT texts of mathematical astronomy from the Seleucid and Parthian periods we find traces of the "ideal year" in the useful unit *tithi* = 1/30 synodic month. I have no doubt that the "ideal year" was simply a continuation of the old accounting practice from the Ur III period and the *tithis* of Late Babylonian astronomy a practical reminiscence of the 360-day year.

Only a few non-astronomical texts from later times prove that the schematic 360-day-year was still considered by the Babylonian scribes. A text with a prayer from the time of Ammizaduga refers to the year (from 20th of Nisan to 20th of Nisan) as having 6 times 60 days and nights.[18] Similarly, some mathematical exercises and coefficients of work are based on the practical approximation that each lunar month was counted as 30 days, independently of its actual length. There has not, however, been found any single administrative text from the later time of the kind which could prove that the schematic 360-day year was used in everyday practice. From the Old Babylonian period and onwards here exists no evidence of such advance planning that gives the ration for the whole year and also the daily ration. Maybe the way of registration and control changed after the Ur III period? We know that Šulgi had introduced major changes into the accounting system (e.g., by forcing through that only the sexagesimal number system be used by calculations). Indeed, E. Robson (1999), p. 171 concludes that "scribal students in the Ur III period were taught to perform their calculations in sexagesimal notation" (see also *ibid* p. 169). This new practice may explain the "missing texts".

We do, however, have evidence of the 360-day year being used in the education of scribes. Some mathematical texts used in scribal schools state a problem and give a model solution. Such texts show how Babylonian mathematics functioned and what pupils were to learn. The subject of many problem texts is clearly connected to bureaucracy and administrative accounting, as is described in the Neo-Assyrian "Examenstext A".[19]

> Do you know multiplication, how to find reciprocals and coefficients, bookkeeping, administrative accounting, how to make all kinds of pay allotments, (can you) divide property and delimit shares of fields?

One interesting problem text, the Old Babylonian mathematical text BM 85196 (xiii), is discussed by Robson (1999), pp. 80–81. The text makes use of the 360-day year. The problem consists in finding how much time it will take one man to carry 6,00 (= 360) sheaves over a long distance: Three hundred and sixty sheaves are arranged in a line, each 5 nindan ($\simeq$ 5 $\times$ 6 metre) apart, so that the furthest is 30,00 nindan ($\simeq$ 10.8 km) and the closest 5 nindan away. A worker has to carry them, one at the time, to the same place. The problem is to find out how long a time it will take the carrier to do it. The text calculates the mean distance to be

$$(30{,}00 + 5) \div 2 = 15{,}02;30$$

and by multiplying it by 6,00 (the number of sheaves), the sum of the distances, over which the sheaves have to be carried, is correctly found to be[20]

$$6{,}00 \times 15{,}02;30 = 1{,}30{,}15{,}00$$

i.e., 324,900 nindan ($\simeq$ 1950 km). This total distance is then divided by 45,00 nindan, which is the daily distance a worker is expected to carry the goods

1,30,15,00 ÷ 45,00 = 2,00;20

Note that the daily distance 45,00 nindan gives only the length over which the sheaves are carried. Therefore the worker going forth and back is expected to walk twice as long, i.e., 1,30,00 nindan (≃ 32.5 km) a day.

The text has shown how to solve the problem and ends with the following solution:

> You will see 2,00;20. He carried here in 4 months and ⅓ day.

This last remark attests evidently the use of the 360-day year: Since 2,00;20 = 120⅓ we see that each of the four months was calculated as 30 days. My point is, that the scribes or pupils working with such texts were aware of the of the practical approximation of 30 days to the lunar month—in other words, they knew about the artificial year of 12 × 30 days.

The Coefficient Lists

From Old Babylonian times and later we have coefficient lists and mathematical problem texts using coefficients. These coefficient lists also indicate that the administrative 360-day year was used in Old Babylonian scribal training. A coefficient, called IGI.GUB in the cuneiform texts, is a parameter which is not given in the statement of a problem but is essential for its solution (for example, 45,00 the daily rate of carrying sheaves). The scribes were supposed to know the coefficients or where to find them in order to solve the problem.

We use π for finding the area of a circle. The Old Babylonian method for finding the area of a circle from its circumference goes as follows: one calculated the square of the circumference and multiplied the result by 0;05. The coefficient 0;05 = 1/12 used for finding the area of the circle was given in many coefficient lists: "0;05, the coefficient of a circle" or "0;05, of the area of a circle". Geometrical coefficients were used for finding areas or volumes while other coefficients were used for Old Babylonian administrative calculations in mathematical problem texts.

Daily working rates were given by coefficients: a worker was expected to carry the load of 540 standardized bricks over 30 nindan in a day, or to demolish the volume of 0;20 sar_v of a wall per day. The working rate for a month was just 30 times the daily working rate. Evidently, these coefficients and the calculations in mathematical texts utilize the artificial year of 360 days, whilst daily life was regulated by the lunisolar calendar.

That these Old Babylonian coefficients and calculations, indeed, go back to the Ur III period has been shown by Robson (1999), p. 160. Here she has published an Ur III text that gives a detailed record of demolition work. A wall of length 22 nindan, average width 2 cubits (= 2 × 1/12 nindan), and height ½ nindan 2 cubit shall be demolished, the total volume being 29 ⅓ sar_v. The text states: "the wages per worker are ⅓ sar_v each". Then it gives the correct answer: "Its workers are 1,28 for 1 day" (29⅓ × 3 = 1,28). The Ur III wage per worker = ⅓ sar_v each, equals the coefficient 0;20 sar_v of wall per day known from Old Babylonian texts. Robson concludes: "In other words, this [Ur III] document provides unequivocal evidence that the daily rate of demolition used in Ur III quantity surveying calculations was later used in OB scribal training".[21]

Analysis of Ur III work rates, Old Babylonian coefficient lists, and Old Babylonian mathematical texts have thus shown that accounting practices involving the artificial year of 12 × 30 days in either administrative or scribal training contexts are found throughout the time from Ur III to the Old Babylonian period. And from this time onwards, we have evidence for the utilization of the artificial 360-day year in "astronomical" texts.

Already during Old Babylonian times, the schematic year of 12 months of 30 days was used for recording astronomical regularities. The Old Babylonian text BM 17175+17284 contains a scheme that connects day length and season (time within the year).[22] The text places the solstices and equinoxes on the 15th of Months XII, III, VI and IX and the day length varies linearly between 2 minas and 4 minas. This text is a forerunner for the "first intercalation scheme" in MUL.APIN. Having thus shown that the "ideal year" of MUL.APIN can be seen as part of a old tradition, originating in the 360 days accounting year from archaic times, I shall briefly mention how this "regular grid" was used in the early formation of astronomical theory.

Within the astronomical schemes of MUL.APIN, the ideal year can again be interpreted as a grid imposed upon and adjusted to nature. This can be seen by the fact that the "first intercalation scheme" indicates some observations aiming at finding some "extra days" which tell how far, for example, the real spring equinox is from full moon (day 15) of Month I. In the ideal case, solstices and equinoxes were assumed to take place in the middle of Month I, IV, VII, and X, and the length of the night was found through linear interpolation between the extreme values, 2 minas on day 15 of Month IV and 4 minas on day 15 of Month X. Therefore, in the ideal case, the duration of day and night would be 3 minas on day 15 in Month I and VII, and the sun would rise straight east. Imagine a grid with bars, representing the time axis with intersections in months: the bars being situated 30 days apart, each of them marking the beginning of a month. In the ideal case, sunrise straight east would happen at full moon (day 15) of Month I, and the schematic length of day and night would be as given in the ideal scheme. In the normal case, however, sunrise straight east did not take place at full moon. The grid, being dependent of the real month, would have to be adjusted—pushed in such a way—that for the specific month in question, the bar would coincide with the time of first visibility of the new crescent.

In this case, sunrise straight east would not take place on day 15; the bars would not be situated symmetrically around this sunrise which tells that day and night will be equal. The shift in days between the ideal place of the grid and its actual place could be observed, and the length of the night at full moon (on day 15) would have to be corrected accordingly. In MUL.APIN there are, indeed, indications for such an interpolation, necessary for finding the actual length of the night at full moon of Month I. In Brack-Bernsen (2005), I have analyzed two schemes of MUL.APIN and argued for the fact that "ideal values" were adjusted. In the so-called "first intercalation scheme", the lengths of day and night were given for the middle of Month I, IV, VII, and IX, together with indications where on the eastern horizon the sun would rise. The "water-clock scheme" was derived from the "first intercalation scheme": the length of the night (given in the "first intercalation scheme") was used as a generating function for finding the "visibility of the moon" which was recorded for the 1st (new crescent) and the 15th (full moon) of 12 ideal months. Equinoxes and solstices do normally not take place at full moon, but some "extra days" away. Therefore the value of visibility was corrected by interpolation according to the shift in night length imposed by the "extra days".[23] That the "ideal year" of 360 days is meant as an approximation to the real lunisolar year becomes also clear, when MUL.APIN writes about intercalations every three years and calculates the effect of such intercalations (Tablet II ii 11–17).

The concepts behind the two schematic calendars (the archaic administrative calendar and the ideal year) are quite similar, and so are their dependences on the real lunar calendar. Both serve as a means of simplifying calculations, and both are coupled to the real lunar calendar and adjusted accordingly. The archaic calendar was adjusted by adding an extra month in tune with the intercalations necessary (on average every three years) within the lunisolar calendar. This means that each lunar month was just calculated as having 30 days independently of its actual length. The "bars" of the grid did always coincide with the beginning of a new lunar month. The same is true for the ideal calendar when used much later in astronomical schemes—the ideal year can be seen as the continuation of the artificial year of accounting. Therefore, it is just the logical consequence, that the daylength (together with all derived quantities) would have to be corrected when the "bars" were displaced with respect to their ideal places symmetrically around solstices and equinoxes.

Conclusions

We have seen how the administrative and the cultic/civil calendars coexisted all way through from the earliest literate time (*ca.* 2600 BC) until *ca.* 300 BC. The astronomical lunisolar calendar regulated life, while the artificial accounting or ideal calendar (which I see as a practical and regular approximation to nature) could be used for calculation.

Acknowledgments

I am grateful to Hermann Hunger for useful references and for his careful reading of the manuscript. I also thank the Deutsche Forschungsgemeinschaft for supporting this work.

Notes

1. For lack of a better expression the word "calendar" is (in the artificial "administrative calendar") applied for denoting the practical way of reckoning time, where each month (independently of its real duration) is approximated by 30 days.
2. See Vaiman (1989) and (1990), where many of Vaiman's earlier papers, written in Russian, have been slightly revised and translated into German.
3. Vaiman (1989 and 1990), Friberg (1978).
4. See Green *et al.* (1987, pp. 117–166) and Nissen *et al.* (1990, pp. 61–65).
5. The Sexagesimal system is attested in the Uruk IV – III Periods (3100 BC – 2900 BC), i.e., in periods much earlier than any secure attestation of the Sumerian Language (see Englund (1988), p. 122). Powell (1972) sees the situation differently. He has argued for the fact that within the Sumerian language number words have a sexagesimal structure and so concludes that the sexagesimal notations of the period must have originated in a spoken Sumerian.
6. See Englund (1988), p. 123, footnote 2.
7. See Englund (1988), p. 123.
8. Englund (1988), p. 123, footnote 3.
9. Englund (1988), pp. 143–147.
10. See Nissen et al. (1990).
11. See Englund (1988), p. 133, footnote 10.
12. See Beaulieu (1993) and Brown (2000), pp. 146–153.
13. See Brack-Bernsen and Hunger (2002).
14. Neugebauer (1955).
15. von Weiher (1988), pp. 198–200.
16. For futher details, see Brack-Bernsen and Steele (2004) and Steele (2006).
17. Another example can be mentioned here: Two sections of the Atypical Astronomical Cuneiform Text E

published by Neugebauer and Sachs (1967) are concerned with lunar latitude. A new analysis of text E has shown how the "ideal year" was utilized for the construction of a zigzag function for the lunar latitude; see Brack-Bernsen and Hunger (2007).

18. See de Meyer, 1982, pp. 274–275. I thank H. Hunger for this reference.
19. See Sjöberg (1975), p. 145, who gives a German translation.
20. See also Friberg (1987–90), p. 578.
21. See also Englund (1988), p. 177 where, commenting of Ur III workload texts, he surmises that a variant of the Ur III workload system may have been in use already in presargonic Lagash.
22. Hunger and Pingree (1988), pp. 163–164.
23. See Brack-Bernsen (2005), pp. 7–13.

References

Beaulieu, P.-A., 1993, "The Impact of Month-lengths on the Neo-Babylonian Cultic Calendar", *Zeitschrift für Assyriologie* 83, 66–87.

Brack-Bernsen, L., 2005, "The 'days in excess' from MUL.APIN. On the 'First Intercalation' and 'Waterclock' Schemes from MUL.APIN", *Centaurus* 47, 1–29.

——, and Hunger, H., 2002, "TU 11, A Collection of Rules for the Prediction of Lunar Phases and of Month Lengths", *SCIAMVS* 3, 3–90.

——, and Hunger, H., 2007, "On the Atypical Astronomical Cuneiform Text E, A Meanvalue Scheme for Predicting Lunar Latitude", *Archiv für Orientforschung*, in press.

——, and Steele, J. M., 2004, "Babylonian Mathemagics: Two Mathematical Astronomical-Astrological Texts", in C. Burnett, J.P. Hogendijk, K. Plofker and M. Yano (eds.), *Studies in the History of the Exact Sciences in Honour of David Pingree* (Leiden, Boston: Brill), 95–125.

Brown, D, 2000, *Mesopotamian Planetary Astronomy-Astrology* (Groningen: Styx).

Damerow, P. and Englund, R. K., 1987, "Die Zahlenzeichensysteme der archaischen Texte aus Uruk", in *Archaische Texte aus Uruk* 2 (Berlin: Gbr. Mann Verlag) 117–166.

Englund R. K., 1988, "Administrative Timekeeping in Ancient Mesopotamia", *Journal of the Economic and Social History of the Orient* 31, 121–185.

——, 1998, "Texts from the late Uruk Period", in *Mesopotamien, Späturuk-Zeit und Frühdynastische Zeit*, Orbis Biblicus et Orientalis 160/1, 15–233.

——, 2001, "Grain Accounting Practices in Archaic Mesopotamia", in J. Høyrup and P. Damerow (eds.), *Changing Views on Ancient Near Eastern Mathematics* (Berlin: Dietrich Reimer Verlag), 1–35

Friberg, J., 1978, *The Early Roots of Babylonian Mathematics* I (Göteborg).

——, Hunger, H. and al-Rawi, F. N. H., 1987–90, "Seed and Reeds, A Metro-Mathematical Topic Text from Late Babylonian Uruk", *Baghdader Mitteilungen* 21, 483–557.

Green, M. W., Nissen, H. J., Damerow, P. and Englund, R. K., 1987, *Zeichenliste der Archaischen Texte aus Uruk. Archaische Texte aus Uruk* 2 (Berlin: Gebr. Mann Verlag).

Hunger, H. and Pingree, D., 1989, *MULAPIN: An Astronomical Compendium in Cuneiform*, Archiv für Orientforschung Beiheft 24 (Horn: Verlag F. Berger).

Lansberger, B., 1949, "Jahreszeiten im Sumerisch-Akkadischen", *Journal of Near Eastern Studies* 8, 248–297.

de Meyer, L., 1982, "Deux prières *ikribu* du temps d'Ammisaduqa", in G. van Driel *et al.* (ed.), *Zikir shumim: Assyriological Studies presented to F. R. Kraus* (Leiden: Brill), 274–275.

Neugebauer, O., 1955, *Astronomical Cuneiform Texts* (London: Lund Humphries).

——, and Sachs, A., 1967, "Some Atypical Astronomical Cuneiform Texts I", *Journal of Cuneiform Studies* 21, 183–218.

Nissen, H. J., Damerow, P. and Englund, R. K., 1990, *Frühe Schrift und Techniken der Wirtschaftsverwaltung im alten Vorderen Orient* (Berlin: Verlag Franzbecker).

Powell, M. A., 1972, "Sumerian Area Measures and the Alleged Decimal Substratum", *Zeitschrift für Assyriologie* 62, 165–221.

Robson, E., 1999, *Mesopotamian Mathematics, 2100–1600 BC, Technical Constants in Bureaucracy and Education*, Oxford Editions of Cuneiform Texts 14 (Oxford: Clarendon Press).

Sallaberger, W., 1993, *Der kultische Kalender der Ur III–Zeit* (Berlin, New York: Walter de Gruyter)

Schmandt-Besserat, D., 1977, "An Archaic Recording System and the Origin of Writing", *Syro-Mesopotamian Studies* 1, 31–70.

Sjöberg, Å. W., 1975, "Der Examenstext A", *Zeitschrift für Assyriologie* 64, 137–176.

Steele, J. M., 2006, "Greek Influence on Babylonian Astronomy?", *Mediterranean Archaeology and Archaeometry* 6/3, 149–156.

Tanret, M., 2004, "What a Difference a Day Made...: On Old Babylonian Month Lengths", *Journal of Cuneiform Studies* 56, 5–12.

Vajman, A. A., 1989, "Beiträge zur Entzifferung der Archaischen Schriften Vorderasiens. Erster Teil", *Baghdader Mitteilungen* 20, 91–138.

——, 1990, "Beiträge zur Entzifferung der Archaischen Schriften Vorderasiens. Zweiter Teil", *Baghdader Mitteilungen* 21, 91–124.

von Weiher, E., 1988, *Spätbabylonische Texte aus Uruk* 3 (Berlin: Gebr. Mann Verlag).

The Astrolabes: Astronomy, Theology, and Chronology

Wayne Horowitz

Introduction

In the Babylonian national epic Enuma Elish, Tablet V lines 1–8, Marduk, the newly crowned King of the Gods, takes it upon himself to arrange the luminaries in the heavens in the wake of his victory over Tiamat at the end of Tablet IV. Here Marduk assigns three stars to each month of the year, and sets the station of his star *Nēberu* alongside the stations of Enlil and Ea to regulate the stars:

Enuma Elish V: 1-8
1. He fashioned the stations for the great gods.
2. The stars, their likeness, he set up, the constellations.

3. He fixed the year, drew the boundary-lines.
4. Set up three stars each for the 12 months.

5. After he drew up the designs of the days of the year.
6. He set fast the station of *Nēberu* ('The Crossing') to fix their bands.

7. So that none would transgress, be neglectful at all,
8. He set the station of Enlil and Ea with it.[1]

These acts of Marduk bring into existence a system by which the starry sky is divided by 12 radii demarking the 12 months of the year; and 3 concentric circles marking the borders between the three stellar paths: the Path of Enlil in the northern part of the sky, the Path of Anu in the central band of the sky, and the Path of Ea in the southern part of the sky. Thus, 36 stellar sectors are established with one sector for each of the 36 stars (see figure 1).

As described, the station of Marduk's own star *Nēberu*, 'The Crossing,' together with the stations of Enlil and Ea, regulates the stars' annual movements. This system is none other than that of a group of cuneiform astronomical texts known to modern Assyriology as 'Astrolabes,' or perhaps better by their native name, 'The Three Stars Each,' *kakkabū* $3^{\text{ta.àm}}$;[2] a name which occurs both in Enuma Elish V 4,[3] and in the astronomical report SAA 8: 19.[4]

This Astrolabe group of texts is at one and the same time both canonical in some ways and non-canonical in others. All texts in the group share the basic principal of Astrolabes: that one star rose in each of the three stellar paths during each of the 12 months of the year, and that these 36 stars thus fixed the months of the annual calendar astronomically in place. For

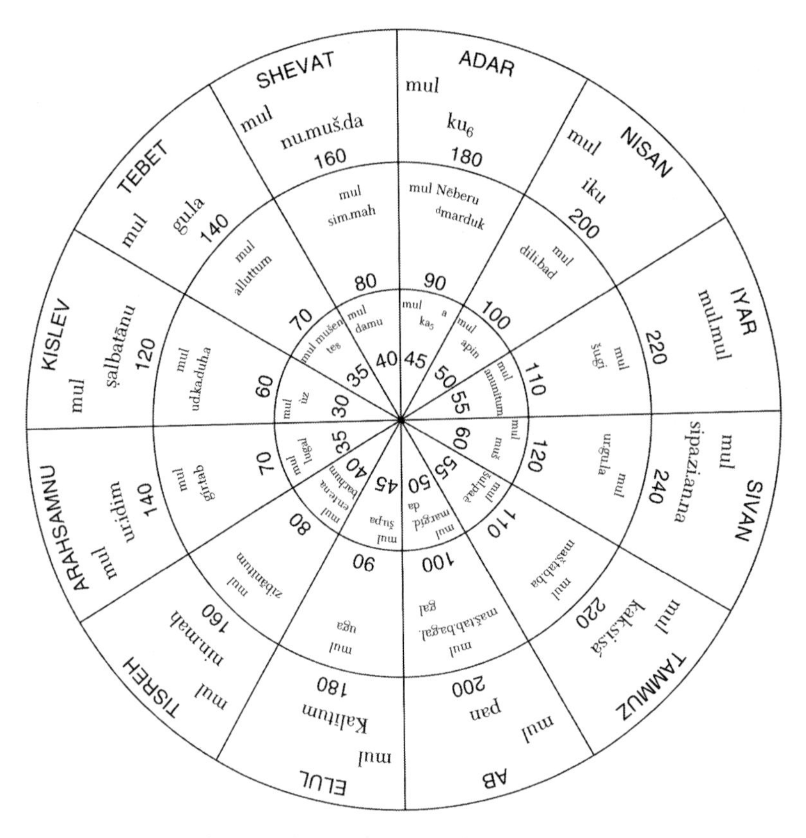

FIGURE 1. The division of the starry sky. (After Horowitx (1998), p. 156.)

instance, the rising of the first star of the Astrolabes, mul iku 'The Field,' (Pegasus), marked the month of Nisan in the Path of Ea. Thus, each of the lunar months of the Mesopotamian calendar, according to the theory of the Astrolabes, was ideally marked not only by its lunar cycle with the first of the month indicated by the new moon, and middle of the month by the full moon etc., but also by the rising of the months' three stars, what I shall call month-stars.

Ideally, at least, the Ea-star seems to have been intended to rise heliacally on the first of the month on the eastern horizon just before dawn a few hours after the appearance of the new moon at dusk on the western horizon. If the month-stars and sequence of new moons were synchronized correctly, this would mean that the lunar calendar was aligned with the stellar cycle. If not, intercalation was necessary to bring the lunar calendar with its regular years of 12 lunar months of 354 days, and leap years of 13 months (383/384 days) back into agreement with the stellar year of 365+ days. Hence, Enuma Elish V 1–8 states that the stars determined the year, just as the Moon-god is commanded by Marduk in Enuma Elish V 11–22 to indicate the days of the month, and the Sun-god in Enuma Elish V 25–46 to demark the watches of day and night.[5]

Yet, despite this agreement on principle, the Astrolabe group is non-canonical in that the scribes never developed a set format for presenting the lists of the 36 month-stars and related materials as is the case, for example, with the Neo-Assyrian period astronomical treatise Mul-Apin that to a large degree supercedes the Astrolabes in the first millennium. Instead, different versions of the Astrolabes circulated with even the most basic element of the Astrolabe tradition, the lists of the 36 month-stars, presented in two distinct and separate formats: circular and list.[6]

Circular and List-Astrolabes

Only two fragmentary examples of circular-Astrolabes are known, both from the Neo-Assyrian period,[7] but a much larger number of Astrolabes written in list format survive.[8] Even here, however, there are wide variations in format and content. For example, the Middle-Assyrian period Astrolabe compendium Alb B (KAV 218) from the 12th century, consists of four elements:

Alb B I: A menology in which 10 months of the year are assigned month-stars.

Alb B II: A star-catalogue listing 36 stars (12 for each of the Paths of Anu, Ea, and Enlil).

Alb B III: A list-Astrolabe with its list of 36 month-stars (1 star for each path per month) in which the repertoire of stars is slightly different from that of Alb B II.

Alb B IV: A list of 36 rising and setting stars in which the stars of Alb B III rise in their path in their assigned month, and then set six months later.[9]

These four elements never occur together in any other known Astrolabe source, although the various elements of Alb B have a few precursors in the Kassite period, and find numerous duplicates and parallels throughout the first millennium. For example, the Star-Catalogue Alb B II is known in a slightly earlier 30 star format—10 stars for each path rather than 12—in two parallels that are separated in time by about a millennium:

1) The Middle Babylonian Nippur tablet HS 1897, which is slightly older than Alb B and dedicated in full to the 30 star-catalogue. This would appear to be a forerunner, so to speak, to the 36 star-catalogue of Alb B Section II.

2) The Hellenistic period Astrolabe compendium, BM 55502, which preserves a duplicate of the list of the 36 rising and setting stars best known from Alb B IV, but also gives a 30 star-catalogue that nearly duplicates that of HS 1897. This is a bit of a surprise since one might have expected the 36 star-version known from Alb B II to have entered the canon rather than the 30 star-version found on HS 1897.[10]

Other sources belonging to the Astrolabe group include omens incorporated into Enuma Anu Enlil Assumed Tablet 51,[11] an esoteric learned commentary (*mukallimtu*),[12] and assorted fragments. Tablets bearing Astrolabes also preserve a version of the mathematical problem text 'The Hilprecht Text,'[13] and materials belonging to a small group of *zipqu*-star texts concerned with the micro-zodiac (*dodekatemoria*).[14] Thus, issues of canonicity, text history, and transmission in regard to the Astrolabe group are unusually complex.

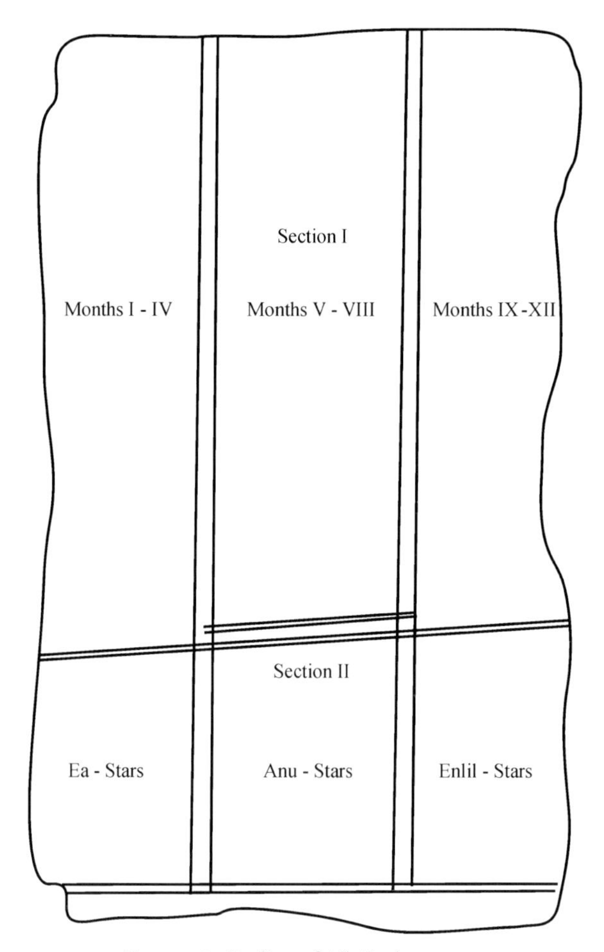

FIGURE 2. Outline of Alb B obv.

Texts in the Astrolabe group date from the Kassite period down into the late period. The earliest texts are the aforementioned 10 Star-Catalogue HS 1897 from Nippur, and a contemporary Kassite period forerunner to the Astrolabe menology (Alb B I) VS 24 120 from Babylon. Numerous sources date to the Neo-Assyrian period including the fragments of circular-Astrolabes. A substantial number of texts date to the late-Babylonian period as well. However, despite this multiplicity of sources, any examination of the Astrolabes and their place in cuneiform civilization must begin with the most detailed and complete of the Astrolabe texts: Astrolabe B.

Astrolabe B

Alb B = KAV 218 = VAT 9416 belongs to the collection of cuneiform tablets which E. Weidner once identified as the 'library of Tiglath-Pileser I.'[15] The tablet is identified by its colophon as being written by the scribe *Marduk-balāssu-ēreš*, son of *Ninurta-uballissu*, the royal scribe, in the eponym year of *Ikkaru*, some time in the early to middle 12th century.[16] The text on KAV 218, as opposed to the tablet, cannot be more precisely dated, but as we

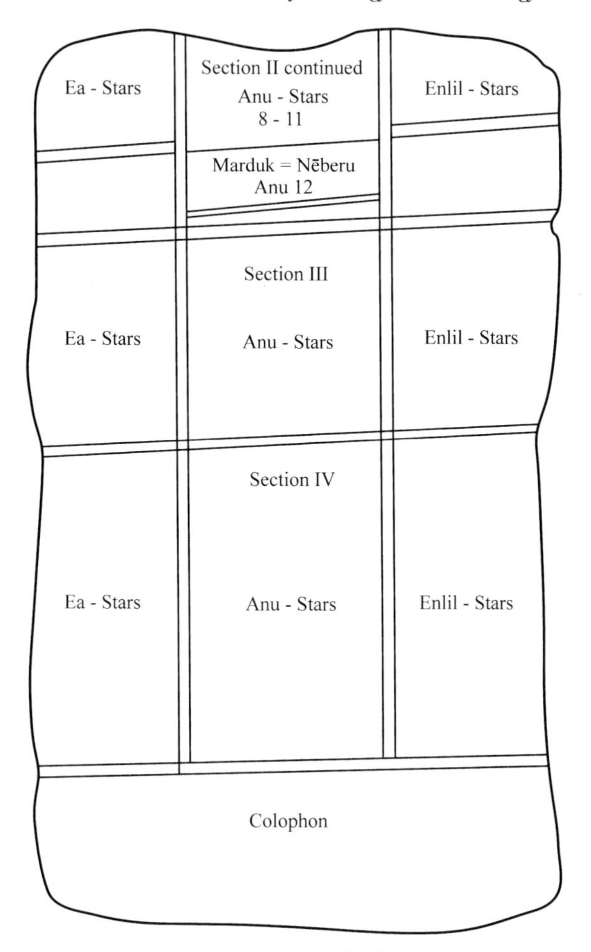

FIGURE 3. Outline of Alb B rev.

shall see below it seems most likely that the contents of Alb B, as we know it, are not much older than KAV 218 itself.

Alb B is inscribed on three columns on both sides of KAV 218 with the text divided into its four sections by double horizontal rulings (see figures 2 and 3). The four sections are, now in more detail:

I) A bilingual menology for the 12 months of the Babylonian year. In each monthly section all lines of the Sumerian version are given first followed by the entire Akkadian version. Among the topics noted in the menology are rites and rituals for the various months, agricultural events, and divinities associated with particular months. In 10 of the 12 months the first item noted in the Sumerian version of the menology is the month-star for that month. For example, in the case of Nisan (Month I), 'The Field' (^{mul}iku = *ikū*, Pegasus) and for Iyar (Month II), mul.mul = *zappu* (Pleiades). The Akkadian version strangely gives only 8 of the 10 month-stars.

II) The Star Catalogue which presents 12 stars for each of the three stellar paths yielding a repertoire of 36 stars. Here Alb B II typically notes the position of each star vis-a-vis those before or after, names the god identified with the star, and occasionally also refers to the colour of the star (red), or particular parts of constellations, for example, the wings of the constellation ^{mul}UD.KA.DUH.A 'The Demon with The Gaping Mouth.'[17] In the Star Catalogue the sequence of stars does not follow the sequence of heliacal risings so these stars are not month-stars. Thus, the repertoire of stars in Alb B II differs from that in Alb III–IV where the stars are month-stars. The most important star in the Star Catalogue Alb B II is that of Marduk, ^{mul}*Nēberu*, 'The Crossing;' this being the 12th and last star in the Path of Anu in Alb B II, and the month-star/rising-star for Adar, the last month of the year, in the Path of Anu in Alb B III-IV. In Alb B II, the discussion of Marduk's star is found at the bottom of the middle column, hanging into open space, below the portions of col. i and iii relating to the 12th and last stars of the Paths of Ea and Enlil. Thus, in Alb B II, Marduk's star seems to be placed after the other 35 stars of the Astrolabe (see figure 3). Given that the annual pattern of risings of stars begins to repeats itself each Nisan, Marduk's stellar position in Alb B II may be understood to indicate that Marduk's star is not only the last star of the old year, but also the first star of the new year.

III) Alb B III is the heart of the Astrolabe as it gives the list of 36 month-stars with the 12 stars in each path listed by month according to the sequence of their presumed heliacal risings.

IV) The list of 36 rising and setting stars in which the stars that rise each month are the month-stars named in Alb B III. 34 of the 36 stars set exactly six months after rising in accordance with a false astronomical theory that non-circumpolar stars rise and set at half year intervals. The two month-stars that do not set in Alb B IV are Venus (Anu 1) which is identified as a planet, and ^{mul}APIN, 'The Plow' (Enlil 1) which is said 'to be present all year long,' i.e. to be a circumpolar star.

It is not certain how, if at all, the information given in the various elements was to be applied to everyday life although one could conceivably have used Alb B as follows. Part I, the menology, could have been used to find correspondences between rising of stars and important events in the annual temple and agricultural calendars. Part II, 'the Star Catalogue' would have allowed the reader to locate important stars in the sky including most of these named in Parts I, III, and IV. Part III and the rising stars of Part IV (in theory at least) would have provided the reader with the ability to identify the month of the year astronomically on the basis of the risings of stars, as well as to intercalate the calendar by observing the date of heliacal risings of stars against the progression of the months of the lunar calendar. Part IV, with its false theory for the setting of stars, would have added no new useful information on its own.

I suggest that Alb B was more than simply an astronomical aid for calendar keeping, but rather a sort of astronomical handbook which had some practical applications, as well as a purely academic function as a statement of astronomical theory with religious overtones; namely that the stars were identified with gods, and that Marduk, whose star is to be found in the pivotal position at the end of the Path of Anu in Alb B II, was the god who regulated the starry sky and so the passage of time.

The History of The Astrolabe Tradition

The history of the Astrolabe tradition can be divided into the era before Alb B and the era after Alb B. Before Alb B we have but two sources, both from the Kassite period, which provide antecedents to Alb B Sections I and II: 1) VS 24 120 from Babylon with its unilingual Sumerian short version of the menology found in Alb B I for only the first six months of the year, without month stars, and 2) HS 1897 from Nippur with its shorter version of the Star Catalogue Alb B II: 10 stars per path for the Paths of Anu, Enlil, and Ea, instead of 12, with some astrological material attached. HS 1897 is particularly important for the history of the Astrolabe tradition in that it, along with the roughly contemporary Prayer to the Gods of the Night KUB 4 47 from Boghazoköi, provides the earliest direct evidence for the division of the starry sky into the three traditional paths of Mesopotamian astronomy.[18]

Beyond the observation that VS 24 120 is an antecedent to Alb B I, and HS 1897 to Alb B II, a further connection can be made between both texts and the menology Alb B I. As noted above, VS 24 120, in addition to offering an abbreviated version of the Sumerian portion of Alb B II menology for only Months I–VI, differs from Alb B I in that the earlier Kassite period text does not assign any month-stars, while the menology Alb B does for 10 of the 12 months. This suggests that Alb B I as we know it is based on at least two separate *vorlage*:

1) A complete form of the menology for all 12 months without month-stars.

2) A list of stars, most likely in the sequence of their heliacal risings, that could have been used to assign stars to at least most months of the year so as complete the menology in the form available in Alb B I.

Here the fact that the menology Alb B I gives month-stars for only 10 of the 12 months would seem to indicate that the editor of Alb B I drew upon a pre-existing list of 10 stars such as those for the Paths of Anu, Enlil, and Ea in HS 1897. It may therefore come as no surprise that the 10 Stars of Ea in HS 1897 provide a near perfect match with the 10 month-stars in the menology Alb B I.[19]

In contrast, there are no precursors or forerunners of any kind to Alb B III–IV, which represent the first surviving lists with 12 month-stars per path giving a total of 36 stars.[20] This would seem to indicate that Alb B III–IV is the work of an editor writing later than the time of the Kassite period tablets HS 1897 and VS 24 120. In fact, I would suggest that the author of what we today know as Alb B wrote his text proximate to the time of the Middle Assyrian tablet KAV 218 itself.

In the first millennium, the four elements brought together by the editor of Alb B go their separate ways. The menology Alb B I exists in a form almost identical to that on Alb B on the Neo-Assyrian tablet Sm. 755+ from Nineveh, but two parallels to the menology found in Enuma Anu Enlil Assumed Tablet 51 are substantially different. The first is closer to Alb B I both in content and form in that it offers bilingual sections for each of the 12 months,[21] but the second gives bilingual sections for 13 months, including an intercalary Adar, and preserves some materials not found in Alb B I or the 12 month menology of EAE 51.[22] This demonstrates that innovation in the Astrolabe tradition continued into the first millennium.

The Star-Catalogue Alb B II in its 36 star format, in contrast, has yet to be found in first millennium although the 30 star version occurs on BM 55502. Nonetheless, it is possible that the 36 star version existed as well for which we as yet have no witness.

The list of 36 month-stars first known from Alb B III is repeated numerous times in different formats and contexts in the first millennium. The circular Astrolabe fragments from the Neo-Assyrian period, when complete, diagrammed this repertoire of stars on planispheres adding monthly values for day and night hours over the course of the year. These values were given along with the star-names in the 36 stellar sectors on the planispheres with the Ea month-star given full value, the Anu month-star 1/2 value, and the Enlil month-star 1/4 value. The full values, 3 at the equinoxes, 4 at the summer solstice, and 2 at the winter solstice, refer to the length of daylight over the course of the year, and are found for the first time centuries before Alb B in the Old Babylonian period.[23] The assignment of these values to the month-stars is a common feature of most first millennium Astrolabe texts, occurring as well in the list-Astrolabe and omens of LBAT 1499, and the *mukallimtu*-commentary to Astrolabes BM 82923. To these sources can be added a partially preserved list of the Astrolabe stars in the late-period astrological work TCL 6 13,[24] and a few fragments.

The fourth and last section of Alb B, the list of rising and setting stars, is known in the late-period from the Astrolabe compendium BM 55502 and its contemporary Rm. 781 which are written in the same format as Alb B IV. A tiny fragment from Assurbanipal's library in the same format proves that the list of rising and setting stars was also known in the Neo-Assyrian period thereby providing a time bridge between Alb B and the late-Babylonian sources. Another late-Babylonian tablet preserves the list of rising and setting stars in a different format. Mention of a system of rising and setting stars at six month intervals such as that espoused in Alb B IV and its parallels, as well as a system of 30 stars akin to that of the star-catalogues HS 1897 and BM 55502, is also to be found in the writings of the first century B.C. Greek historian Diodorus Siculus in the context of his discussion of Babylonian astronomy.[25]

Other materials relating to the Astrolabe group in the first millennium are CT 33 9, a Neo-Assyrian list of 36 stars that utilizes up-dated criteria to assign stars to the Paths of Anu, Enlil, and Ea, and Nv. 10 from Nineveh, now in Istanbul, which V. Donbaz and J. Koch identify as an Astrolabe "of the third generation."[26] A list of 36 stars of Elam, Akkad, and Amurru in The Great Star List seems to be related directly to the Astrolabe group by a fragment of a commentary (*mukallimtu*) to Enuma Anu Enlil from Nineveh.[27] Finally, the aforementioned version of 'the Hilprecht Text,' a mathematical problem text involving distances between stars, is preserved on the reverse of the circular Astrolabe Sm. 162. This provides yet another link between the Astrolabe group and the second millennium. 'The Hilprecht Text' is known earliest from the Kassite Period tablet HS 245 from Nippur.

The separate histories of the four elements of Alb B in the first millennium demonstrate without the shadow of a doubt that Alb B was not admitted as a single piece into the cuneiform canon. Instead, it would appear that Alb B itself was understood even by its own editors to be composed of independent pieces that were joined together. Hence, perhaps, the double horizontal rulings dividing section from section on KAV 218. If so, it is fair to hypothesize that the elements of Alb B were brought together proximate to the time of the writing of the tablet KAV 218 in the 12th century, and that the editor of Alb B (or its immediate source) may have himself added a substantial amount of new material to at least parts of his text. This supposition is bourne out by theological allusions in Alb B to the Mardukian astronomical theory espoused later in Enuma Elish V 1–8.

Astrolabe B and Enuma Elish

In 1964 my teacher Prof. W.G. Lambert of Birmingham University published an article "The Reign of Nebuchadnezzar I, A Turning Point in the History of Ancient Mesopotamian Religion".[28] In this article, Lambert argues that Enuma Elish was composed around the time of Nebuchadnezzar's summer victory over Elam to commemorate the return of Marduk's statue home to Esagil at the very end of the 12th century.[29] This *ca.* 1100 date for the composition of Enuma Elish is now commonly accepted, and it is also generally agreed that this too is the approximate date for the composition of the series Tintir which exalts Marduk's city Babylon.[30] If so, the colophon of KAV 218 demonstrates that the date of our copy of Alb B is also roughly contemporary to that of Enuma Elish, with Alb B on KAV 218 predating Enuma Elish at most by a few generations. Further, the scribes of both KAV 218 and Enuma Elish are both Babylonian: Enuma Elish by a learned man/men of Marduk, presumably at Babylon, and KAV 218 by a Babylonian scribe bearing the name of Marduk, *Marduk-balāssu-ēreš* son of *Ninurta-uballissu*, serving the Assyrian king at Assur. Thus, Enuma Elish and Alb B came to share a common astronomy with Enuma Elish V 1–8 referring to the 36 stars of the Astrolabes. Yet, the common basis of the two works goes far beyond their mere agreement on the importance of the 36 month-star system that stands at the heart of the astronomical theory of Astrolabes, for even the physical format of KAV 218 echoes the teachings of Enuma Elish.

Enuma Elish V 5–8 teaches that the stations (*manzazu*) of Enlil and Ea are set together with the station of Marduk so as to regulate the other stars—again:

Enuma Elish V 5–8
5. After he (Marduk) drew up the designs of the days of the year.
6. He set fast the station of *Nēberu* ('The Crossing') to fix their bands.

7. So that none would transgress, be neglectful at all,
8. He set the station of Enlil and Ea with it.

I suggest that these lines refer to Marduk's star 'The Crossing' and stars identified with Ea and Enlil in Alb B II:

Ea: ^{mul}KU$_6$, 'The Fish,' the last star in The Path of Ea in Alb II and the Ea-star for Adar (Month XII) in Alb B III. (Alb B II i rev. 9 (Ea 12): "['T]he Fish', Ea.")

Enlil: ^{mul}APIN, 'The Plow,' the first star in The Path of Enlil in Alb B II, and the Enlil-star for Nisan (Month I) in Alb B III. (Alb B II iii 1–3 (Enlil 1): "'The Plow' which stands at the rising of the eastwind in front of 'The Wagon', this star is Enlil who determine[s] the destinies of the land.)

Alb B III	Path of Ea	Path of Anu	Path of Enlil
Month I	'The Field'	Venus	'The Plow'
Month XII	['The Fi]sh'	Marduk	'The Fox'

On the tablet itself, the entry for 'The Fish' stands above and to the left of *Nēberu,* with the entry for Marduk's star extending into open space below. Immediately below the entry for Marduk's star, and to its right is the listing for 'The Plow' as the Enlil-star for Nisan in the first line of Alb B III:

God	Star	Path	Month
Ea	'The Fish'	Ea	XII (Adar)
Marduk	'The Crossing'	Anu	XII (Adar)
Enlil	'The Plow'	Enlil	I (Nisan)

Thus, the entry for Marduk's star *Nēberu,* hanging down at the end of the star-catalogue Alb B II, serves as a sort of bridge between the main body of the star-catalogue which precedes it, and the list of month-stars which follows.[31] Thus, Marduk's star *Nēberu* may be imagined to divide the stars of the outgoing old year from those of the incoming new year in line with the description of Marduk's star in the text of the Star Catalogue Alb B II ii rev. 10–13:

> Alb B II ii rev. 10-13
> The red star which stands at the rising of the southwind after the gods of the night have completed (their courses) dividing the heavens, this star is 'The Crossing,' Marduk.

This same point is also made in Ee VII 124–127 where the 49th of the 50 names of Marduk is none other than *Nēberu*:

> Ee VII 124–127
> Let *Nēberu* ('The Crossing') be the holder of the crossing of heaven and earth
> so that they cannot cross above and below, but must wait for him.
> *Nēberu* ('The Crossing') is his star which he made appear in the heavens,
> He is the one who holds 'the turning point,' they must look to him.

Here we may imagine that the stars cannot cross from above to below, in terms of the physical format of Alb B from the old year above to the new year below, without Marduk's permission. As such, I would submit that this is the idea which lies behind the name of Marduk's star, 'The Crossing,' i.e. crossing from the old year to the new (see figure 4).

In terms of Enuma Elish and Astrolabes, the above interpretation explains how Marduk, together with the stations of Enlil and Ea, regulates the year: Enlil's star 'The Plow' regulates Enlil's path; Ea's star 'The Fish' regulates Ea's path; while Marduk's star *Nēberu* regulates the entire system from his position at the very end of the central band of the sky, the Path of Anu. Thus, it would seem that the author(s) of Enuma Elish, writing just a few generations after the scribe of Alb B, knew Alb B, or a text very similar to Alb B.

Conclusion

Marduk's star is placed in the pivotal position of the Alb B as a whole, as the text is inscribed on the tablet KAV 218. Above the star-catalogue's entry for ^{mul}*Nēberu* is half of Alb B (all of Alb B I and the remaining 35 stars of Alb B II), while below is the second half of Alb B (all

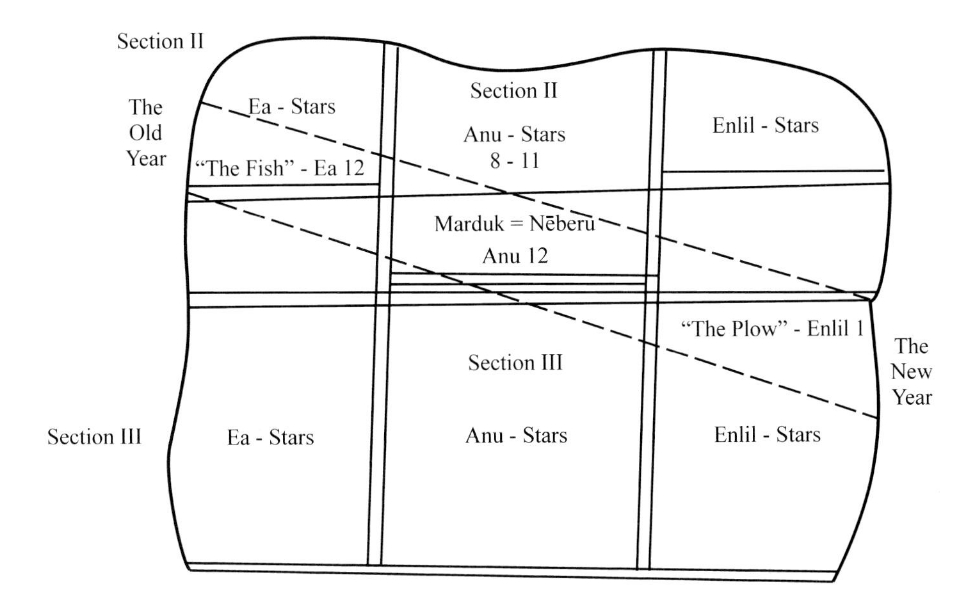

FIGURE 4. The Crossing from the old year to the new year.

of Alb B III–IV). Thus, Marduk occupies a central position in Alb B not only in terms of the astronomy of the text, but also in terms of the tablet's physical format.

This central position of Marduk in Alb B, between above and below, echoes Marduk's role as the new King of the Heaven and Earth in Enuma Elish where Marduk governs the sky by means of his star. Thus, it would appear that Alb B, in the form we know it from the late second millennium just before the date of the composition of Enuma Elish, must have been composed by a man of Marduk who knew the astronomical conventions of his time, and used them in Alb B with the intent of demonstrating that Marduk's star ruled as king over the stars in heaven just as Marduk himself ruled on earth.

Finally, given the affinities between Alb B and Enuma Elish, the date of KAV 218, as established in its colophon, may provide independent confirmation for the *circa* 1100 date of the composition Enuma Elish, while the close ties between the astronomical theory of Alb B and Enuma Elish demonstrate that Alb B is not only an astronomical treatise, a scientific text, but moreover a religious theological work of the highest importance which extolls the central message of Enuma Elish: *Marduk-ma šar* - Marduk is King![32]

Notes

1. Translation adopted from Horowitz (1998) 114–115.
2. The term "Astrolabe," as it is used in Assyriology, does not refer to the antique instrument of this same name, but instead to the group of Mesopotamian cuneiform astronomical texts discussed in this paper. Below I present findings from the author's ongoing research with full publication expected soon in Astrolabes and Related Texts. In the meantime, for editions of some texts in the Astrolabe group see

Casaburi (2003). For discussion of the group see Hunger-Pingree (1999) 50–57; Horowitz (1998) 154–166; van der Waerden (1974) 64–67; and earlier Weidner (1915) 62–102. Assyriological abbreviations are as in CAD (*The Chicago Assyrian Dictionary*).

3. 12 *arhū kakkabū* 3$^{\text{ta.àm}}$ *ušziz.*
4. Previously Parpola (1983) no. 319 and Thompson Rep. 152.
5. Unfortunately much of the solar section is lost due to breaks in the source material.
6. I shall refer to these as circular-Astrolabes and list-Astrolabes.
7. Sm. 162 (CT 33 11) and K. 14943 + 81-7-27, 94 (+) 83-1-18, 608 (CT 33 12).
8. For partial lists of Astrolabe sources see Horowitz (1998) 155 and Casaburi (2003) 25–27. A fully updated list will appear in Astrolabes and Related Texts.
9. This model is of course false. Even the editors of Alb B somewhat deviate from their model by indicating that Venus is a planet, and that ^{mul}APIN, 'The Plow,' is a circumpolar star.
10. For HS 1897 and BM 55502 see Horowitz and Oelsner (1997–1998).
11. See Pingree and Reiner (1981) 52–69.
12. BM 82923 for which see Hunger and Walker (1977).
13. HS 245 and parallels. See for now Horowitz (1998) 177, 179–182. A new edition of HS 245 by J. Oelsner is in press with Archiv für Orientforschung.
14. LBAT 1499 Sections 5-6 recently edited in Rochberg (2004).
15. Weidner (1952–53).
16. See Freydank (1991) 94–97, 140–141. For this dating see also Hunger-Pingree (1999) 51 and Pedersén (1998) 83.
17. Hunger-Pingree (1999) 274: Panther = Cygnus, Lacerta, and parts of Cassiopeia and Cepheus.
18. See Horowitz (1998) 158–159.
19. See the tables in Horowitz and Oelsner (1997–1998) 181–182.
20. It was suggested at one time that the lists of 36 Elam, Akkad, and Amurru-stars found in The Great Star List were forerunners to the Astrolabe lists, but this idea must now be abandoned given the discovery of the 30 star precursor in HS 1897.
21. Pingree and Reiner (1981) 62 X 24–36.
22. Pingree and Reiner (1981) 63 X 37–50. Both Çağirgan (1984) and Casaburi (2003) attempt to integrate the different versions of the menology into a single composite edition. Editions of the different versions will be presented separately in *Astrolabes and Related Texts.*
23. See BM 17175+ published in Hunger and Pingree (1989) 163–164 and discussed in Hunger-Pingree (1999) 50. Alternatively read the numbers as sexagesimal 3,0 (180), 4,0 (240), and 2,0 (120) as on the diagram in figure 1.
24. See Rochberg-Halton (1987) 218.
25. Diodorus Siculus II 30. 6–7 (Loeb Classical Library Diodorus Siculus II pp. 450–453). See Horowitz and Oelsner (1997–1998) 183–184.
26. For CT 33 9 see now Horowitz (1998) 174. For Nv. 10 see Donbaz and Koch (1995) with critical review in Hunger and Pingree (1999) 55–57.
27. See Horowitz (1998) 175–177. For the lists of Elam, Akkad, and Amurru-stars see Koch-Westenholz (1995) 196–199.
28. Lambert (1964).
29. More recently see also Lambert (1992) and the texts presented in translation in Foster (1993) 290-303.
30. For the common ideology and date of Enuma Elish and Tintir see George (1992) 5–7 and slightly later George (1997), both with further bibliography cited.
31. One obtains the same results if one repeats the sequence of Astrolabe B II stars on into the coming year beginning again with the stars for Nisan in Alb B II 1. Here too the entry for 'The Plow' would stand below and to the right of the entry for Marduk's star.
32. Enuma Elish IV 28.

References

Cağirgan, G., 1984, "Three more duplicates to Astrolabe B," *Belleten* 48, 399–416.

Casaburi, M. C., 2003, *Tre-stelle-per-ciascun(-mese): L'Astrolabio B: edizione filologica*, Università Degli Studi di Napoli "L'Orientale", Napoli.

Donbaz, V. and Koch, J., 1995, "Ein Astrolab der dritten Generation: Nv. 10," *Journal of Cuneiform Studies* 47, 63–84.

Foster, B., 1993, *Before The Muses, An Anthology of Akkadian Literature*, CDL Press, Bethesda.

Freydank, H., 1991, *Beiträge zur mittleassyrischen Chronologie und Geshichte*, Akademie Verlag, Berlin.

George, A. R., 1992, *Babylonian Topographical Texts*, Uitgeverij Peeters, Leuven.

——, 1997, "'Bond of The Lands': Babylon, The Cosmic Capital", in G. Wilhelm (ed.), *Die Orientalische Stadt: Kontinuität, Wandel, Bruch*, SDV Saarbrhcker Druckerei und Verlag, Saarbrhcken, 125–145.

Horowitz W., 1998, *Mesopotamian Cosmic Geography*, Eisenbrauns,Winona Lake.

—— and Oelsner J., 1997–1998, "The 30 Star-Catalogue HS 1897 and The Late Parallel BM 55502," *Archiv für Orientforschung* 44–45, 176–185.

Hunger H. and Pingree, D., 1989 *Mul-Apin, An Astronomical Compendium in Cuneiform*, Berger & Söhne, Horn.

—— and Pingree, D., 1999, *Astral Sciences in Mesopotamia*, Brill, Leiden.

—— and Walker, C. B. F., 1977, "Zwölfmaldrei," *Mitteilungen der Deutschen Orient-Gesellschaft* 109, 27–34.

Koch-Westenholz, U., 1995, *Mesopotamian Astrology, An Introduction to Babylonian and Assyrian Celestial Divination*, Museum Tusculanum Press, Copenhagen.

Lambert, W. G., 1964, "The Reign of Nebuchadnezzar I, A Turning Point in the History of Ancient Mesopotamian Religion," in W.S. McCullough (ed.), *The Seed of Wisdom: Essays in Honour of T.J. Meek*, University of Toronto Press, Toronto, 3–11.

——, 1992, "Nippur in Ancient Ideology," in M. Ellis (ed.), *Nippur at the Centennial*, The University Museum, Philadelphia, 119–126.

Parpola, S., 1983, *Letters from Assyrian Scholars to the Kings Esarhaddon and Assurbanipal*, Butzon & Bercker Kevelaer, Neukirchen-Vluyn.

Pedersén, O., 1998, *Archives and Libraries in the Ancient Near East 1500-300 B.C.*, CDL Press, Bethesda.

Reiner, E. and Pingree, D., 1981, *Babylonian Planetary Omens 2, Enūma Anu Enlil, Tablets 50-51*, Undena, Malibu.

Rochberg-Halton F., 1987, "TCL 6 13: Mixed Traditions in Late Babylonian Astrology", *Zeitschrift für Assyriologie* 77, 207–228.

Rochberg, F., 2004, "A Babylonian Rising-Times Scheme in Non-Tabular Astronomical Texts", in C. Burnett et al. (eds.), *Studies in The History of The Exact Science in Honour of David Pingree*, Brill, Leiden, 56–94.

van der Waerden, B. L., 1974 *Science Awakening II: The Birth of Astronomy*, Noordhoff International Publishing, Leiden.

Weidner, E., 1915, *Handbuch der Babylonischen Astronomie*, J.C. Hinrichs'sche Buchhandlung, Leipzig.

——, 1952–1953, "Die Bibliothek Tiglathpilesers I.", *Archiv für Orientforschung* 16, 197–215.

Calendars, Intercalations and Year-Lengths in Mesopotamian Astronomy

John P. Britton

The three topics addressed here are loosely related, but all concern developments in Mesopotamia. The first is calendars, whose evolution, at least for astronomical purposes was fully accomplished by the end of the second millennium BC. The second is intercalation practices, about which we know very little before –750, and whose development was frozen early in the 5th century BC when the 19-year cycle became the basis for the civil calendar. The last is year lengths, estimates of which progressively improved, paralleling the increase in astronomical knowledge from the 7th to the 2nd centuries BC, when details of Mesopotamian astronomy passed to Hellenistic astronomers.

I. Calendars

Three types of calendars are encountered in cuneiform texts from Mesopotamia: civil, administrative, and schematic.

Civil Calendar

In the civil calendar the day began at sunset and equaled the interval from one sunset to the next. A new month began 30 days later, unless the new lunar crescent was seen or thought to be visible on the preceding day, in which case the day was said to be "turned back", and the new month began on day 30 of the old, making the old month 29 instead of 30 days long.[1] Normal years consisted of 12 months, but linkage with the seasonal year was maintained by the intermittent intercalation of extra or "diri" months, usually an extra month XII, but frequently a second month VI and rarely, but occasionally, some other month.

The month names encountered in astronomical texts derive from the 3rd millennium Nippur calendar, shown in Figure 1. A majority of these names relate at least loosely to seasonal events, while the rest describe cultic festivals specific to Nippur. The closest linkage to the seasonal calendar was še-kin-ku$_5$, the month of the barley harvest. This was month XII in the Nippur calendar, but month I in the older Ur III calendar, where it was associated with the celebration of the spring equinox. Thus at least in the south the barley harvest seems to have taken place around the time of the spring equinox, a correspondence which was later incorporated into the Old Babylonian version of the schematic calendar.

The Nippur calendar was adopted throughout the south by the end of the 3rd millennium and was extended to Babylon probably by Samsuiluna around –1780. Samsuiluna also appears to have added the eclectic assemblage of semitic names from the local calendars of subject cities, which thereafter became part of the calendaric tradition, but had neither role

I ^{iti}**bara$_2$**-zag-gar:

month [the proxies of the gods are] placed besides the **throne** (of Enlil?);
zag-mu (edge of the year) New Years festival.

II ^{iti}[ezem-]**gu$_4$**-si-su$_3$:

month the horned **oxen** marched forth;
gusisu festival of the plow

III ^{iti}**sig$_4$**-u$_5$-šub-ba-ga$_2$-gar:

month the **brick** is placed in the mold

IV ^{iti}**šu**-numun:

month of **preparing** [causative: šu] for seed;
šunumun festival

V ^{iti}ne-**IZI**-gar:

the month when the **braziers** [IZI] are lit
ezem-ne-IZI-gar festival for the spirits of the dead who come forth

VI ^{iti}**kin**-^{d}Inanna:

month of the **rectification**(?) of Ishtar

VII ^{iti}**du$_6$**-ku$_3$:

month of the Sacred Mound (dwelling of Enlil's ancestors in Nippur) *duku* festival
focused on communicating with Enlil's ancestors

VIII ${}^{iti(giš)}$**apin**-du$_8$-a:

month the **plow** is let go (i.e. put away)

IX ^{iti}**gan**-gan-(mu)-e$_3$:

month when the **clouds**(?) [gan=min=^{d}Adad] come out

X ^{iti}**ab**-e$_3$:

month of the **father** (commemorating Shulgi)

XI ^{iti}ud$_2$-duru$_5$ (= **ZIZ$_2$**.A):

month (for?) **emmer**

XII ^{iti}**še**-kin-ku$_5$:

month to work with (i.e. harvest) **barley**; *šekinku* festival
[= month I at Ur ~ **spring equinox**; *akiti-šekinku* festival]

XII$_2$ ^{iti}**diri**-še-kin-ku$_5$:

extra še-kin-ku$_5$ (the intercalary month at Nippur)

FIGURE 1. Pre-Sargonic (< –2326) Nippur Calendar (Source: Cohen (1993), 78–124).

nor significance in astronomical contexts. Assyria maintained its own distinctive and a still poorly understood local calendar until around –1100, when a calendar reform during the reign of Tiglath-Pileser I (–1113 to –1075) adopted the month names of the now Standard Mesopotamian Calendar elsewhere, but began the civil year a month earlier than in the Old Babylonian tradition, so that the spring equinox fell in month I instead of month XII named for the barley harvest. Thereafter all of the civil months encountered in the astronomical literature share the same Sumerian names, although undergoing successive abbreviations in orthography as shown in Figure 2.

Month	Nippur (< –2300) Isin & South (ca. –2050) Babylon (ca –1780) Assyria	Akkadian (ca –1780)	NA/NB	LBAT
I	^{iti}**bara$_2$**-zag-gar	*ni-sa-an-nu* (<nisag = first)	**bar(a)$_2$**	**bar**
II	^{iti}[ezem-]**gu$_4$-si-su$_4$**	*a-a-ru* (Sippar XII, Emar,...)	**gu$_4$**	
III	^{iti}**sig$_4$**-u$_5$-šub-ba-ga$_2$-gar	*si-ma-nu* (lex. only)	**sig$_4$**	
IV	^{iti}**šu**-numum-(na)	*du-u$_2$-zu* (Sippar, ...)	**šu**	
V	^{iti}ne-**IZI**-gar	*a-bu* (Susa, Sippar, ...)	**IZI**	
VI	^{iti}**kin**-^{d}Inanna	*u$_2$-lu-lu* (Sippar?, Nuzi)	**kin**	
VI$_2$				**kin-2-kam, kin-a**
VII	^{iti}**du$_6$**-ku$_3$	*taš-ri-tu$_3$* (Emar?)	**du$_6$**	
VIII	$^{\text{iti(giš)}}$**apin**-du$_8$-a	*a-ra-aḫ-sam-na* (O. Pers.)	**apin**	
IX	^{iti}**gan**-gan-(mu)-e$_3$	*ki-si-li-mi* (non-Sem.)	**gan**	
XI	^{iti}ud$_2$-diru$_5$ (=**ZIZ$_2$**.A)	*ša-ba-tu* (Susa XII)	**ZIZ$_2$**	
XII	^{iti}**še**-kin-ku$_5$	*ad-da-ru* (Susa I)	**še**	
XII$_2$	^{iti}**diri-še**-kin-ku$_5$	*ar$_2$-ḫu at-ru ša$_2$*-MIN	**diri-še**	**dir, še dir**

FIGURE 2. Calendaric Orthography (Source: Cohen (1993), 297ff).

Administrative Calendar

The administrative calendar was perhaps more accurately an accounting convention than an actual calendar, but nevertheless appears to have been a conceptual bridge between the civil and schematic calendars. For administrative convenience, and especially in accounting for labour and rations over extended intervals, it was assumed in the 4th and 3rd millennia that all months contained 30 days, so that a 12-month year was credited with 360 days and a 13-month year with 390.[2] The procedure slightly overestimated the amounts which would have been actually provided, but greatly simplified the administrative management of large scale projects. [In North America a similar system remained in use for bank interest calculations until finally abandoned during the 1970s when electronic calculators became ubiquitous.]

The Schematic Calendar

At some point, possibly towards the end of the 3rd millennium, but certainly by early in the 2nd, the administrative calendar with its constant day-count, but variable month-count, was further abstracted into a schematic calendar, devoid of intercalations, in which the seasonal year consisted of 12 schematic "months", each consisting of 30 schematic "days", and thus 360 such "days" in all. The implied correspondence, 1 circuit = 360 units, was extended to the actual day, where it became linked to the Old Babylonian metrology of linear units (Figure 3) through the correspondence that the largest standard unit, 1 kaš-gíd or "long

Name	Sumerian	Written Abbreviation	Relative Size	Metric	Time	Schematic Calendar	Arc
barleycorn (s)	še	še	1 s	3 mm			
finger (f)	su-si	si, u	6 s	17 mm			
cubit (c)	kùš	kùš	30 f	50 cm			
rod (n)	nindan	gar	12 c	6 m	4 sec		0;1 deg
60 rods (u)	UŠ (giš)	UŠ (giš)	60 n	360 m	4 min	1 "day"	1 deg
stage (b)	danna	kaš-gìd	30 u	**10.8 km**	**2 hrs**	1 "month"	30 deg
circuit			12 b		1 day	1 year	360 deg

FIGURE 3. Pre-Sargonic and OB Metrology.

stage", equal to $30^{uš}$ or approximately 11 km, required roughly 1/12th of a day to traverse (at a brisk pace). As a result there emerged a consistent metrology which extended the idealized schematic calendar first to the actual day and then to the concept of circuit and circle:

Schematic Calendar

1 "Month" = 30 "Days"
1 Year = 12 "Months" = 360 "Days"
1 Day ~ 12 Stages = 360 UŠ (time-degrees)
1 Circuit = 12 signs = 360 degrees

Cardinal Phenomena

	Ur III	OB, EAE 14	MUL.APIN
VE	I (ŠE)	XII (ŠE) 15	I (BARA) 15
SS	[IV]	III 15	IV 15
Γ-Sirius			IV 15!*
FE	VII	VI 15	VII 15
WS	[X]	IX 15	X 15

*IV 26 or 27 would have been approximately correct for –1100/–1000
Γ-Sirius would have occurred on the summer solstice around –1900.

During the 2nd millennium, cuneiform texts dealing with astronomical subjects almost exclusively reflect this schematic calendar. The earliest attestation is an Old Babylonian text (BM 17175), in which the cardinal phenomena are equidistantly spaced and located at the midpoints of months XII, III, VI, and IX, which means simply that in a properly regulated calendar the cardinal phenomenon should take place in these months. In particular the spring equinox was placed in the month of the "barley harvest", še-kin-ku_5, as in the old calendars of Nippur and Ur, which was month XII in the Old Babylonian calendar. The same scheme is reflected in Tablet 14 of the omen series, Enūma Anu Enlil (EAE) accompanied by a division of the day into $360^{uš}$, as well as in other texts, probably also of Old Babylonian (Old Babylonian) origin.

A modified version of this schematic calendar also dominates MUL.APIN, an Assyrian astronomical compendium composed around the end of the 2nd millennium, and evidently after Tiglath-Pileser's calendar reform. Here the dates of the cardinal phenomena are shifted by one month with the spring equinox is placed in month I instead of month XII as in the OB version. Consequently according to the Assyrian convention, the calendar year began one month earlier in the seasonal year than in the earlier tradition. Stated differently, in MUL.APIN and related Neo-Assyrian texts the equinoxes and solstices remain equally spaced, but fall one month later in the schematic calendar at the mid-points of "months" I, IV, VII and X. As we shall see, during the second quarter of the 1st millennium, the Babylonians gradually restored the Old Babylonian tradition, even as the schematic calendar gave way to the civil calendar in astronomical literature and practice.

II. Intercalations

A perfectly regulated calendar would have required intercalations every 2.7 years on average, or in practice at intervals of 3 or 2 years with 3-year intervals predominating. Figure 4 shows some simple ways of combining 2 and 3-year intervals between intercalations and the cumulative error in a multiple of the resulting period approximating 30 years. Thus in the first line intercalations every 3 years with no 2-year intervals result in an error after 30 years of roughly a month, by which amount the rising of Sirius would occur later in the calendar than expected. Alternating 2 and 3-year intervals, as in the second line, results in a 5-year period whose error is nearly as large but in the opposite direction. Combining a single 2-year interval with 2 or 3 3-year intervals as in lines 3 and 4, leads to 8 and 11 year periods, whose errors in slightly more than 30 years are reduced respectively to smaller but still noticeable -6 and +5 days. Again these are in opposite directions, so some combination of these two should yield an accurate result. As it happens simply combining these two into a 19-year cycle is the most accurate of the proximate alternatives, but it is doubtful that its accuracy could be distinguished from that of the 27 and 30-year cycles without reliable records spanning 60 years or so.

The errors of the 3- and 5-year intercalation periods would have been inescapably noticeable to anyone concerned with the issue, and even the 8- and 11-year cycles would have produced sensible errors in a very few repetitions. Thus for anyone who cared, very little empirical evidence would have led to the 19-, 27-, and 30-year period relations as the best candidates for an accurate intercalation cycle.

From actual records we find instances of intercalations at intervals of 2 and 3 years from the 3rd millennium on, suggesting that it was known from early times that 3 years was too long an interval between intercalations to maintain a calendar in phase with the seasonal year. On the other hand we also find instances of successive intercalary years and no evidence of any systematic practice. Indeed during the reigns of Hammurapi and Ammiditana intercalations in 4 successive years are attested, corresponding to a shift in the beginning of the civil calendar by roughly 75 days.[3]

The earliest discussion of intercalation practices appears in MUL.APIN, which states that an intercalary month is to be added every 3 years, so that 3 years = 37 months, and 1 year = 12⅓ months. Clearly these are actual rather than schematic months, and if we assume 36 months of 29½ days and a 37th month of 30 days, it follows that 3 years contain 1092 days, and thus that 1 year = 364 days.

I'll say more about this in discussing year lengths. For the moment, however, it suffices to

# of Intervals beween 'I's		'I's (extra	Period Relation					Error (days late/early (+/–)	
3 years	2 years	months)	M	=	Y	M/Y	Years	sid.	trop.
1	0	1	37		3	12;20	30	+31	+31
1	1	2	62		5	12;24	30	–28	–28
2	1	3	99		8	12;22,30	32	–6	–6
3	1	4	136		11	12;21,49,..	33	+5	+5
5	2	7	235		19	12;22,6,18..	38	+0.4	–0.2
7	3	10	334		27	12;22,13,20..	27	–1.3	–1.7
8	3	11	371		30	12;22	30	+1.8	+1.4
	Modern					12;22,7,30.. (sid.)			
						12;22,5,47.. (trop.)			

FIGURE 4. Intercalary Period Relations.

note that MUL.APIN also describes various rules for determining when to add an extra month, which would be unnecessary if the 3-year rule were taken seriously, and which imply that more frequent intercalations were needed to maintain a consistent calendar, but also that no general scheme for such intercalations yet existed. Essentially these intercalation rules amount to guides for determining when a 2-year intercalation was required, absent which an intercalation every 3 years would be assumed.

Neo-Assyrian Period (–750 to –625)

Moving from MUL.APIN into the 1st millennium, we have almost no evidence of actual intercalation practices until the middle of the 8th century, when eclipse records begin to furnish direct evidence of intercalations, and also allow us to determine the number of intercalations during intervals when their precise character and distribution is not otherwise known. Evidence becomes less spotty with the reign of Esarhaddon early in the 7th century, and from the beginning of Šamaš-šum-ukin's reign in –666 onwards, the record of intercalations is nearly complete.

Figure 5 shows the intercalations attested and inferred during the period of Neo-Assyrian dominance to the start of Neo-Babylonian hegemony with the reign of Nabopolassar. Years with an extra month XII or VI are marked respectively with letters "A" and "U", with capital letters denoting attested intercalations and lower case letters inferred intercalations, whose numbers, if not precise nature or date, are confirmed by eclipse records. While the record is far from complete before –685, there is no evidence of intercalations at other than 2 or 3 year intervals, in marked contrast to the irregular practice evidenced during the Old Babylonian period. Here I have arranged the years—somewhat arbitrarily in the early years—in columns of 5, 8 or 19 years so that each column begins with the civil year in that column which starts earliest in the seasonal year. Immediately above the first year in each column is the longitude of the sun at the beginning of that year, expressed as degrees from the vernal point, which exhibits an irregular but persistent shrinkage with time. Thus in –746 the civil year began with the sun –35° short of the vernal point (extremely early even for the Assyrian calendar), while by –628 this distance had decreased to –19°.

One anomaly bears comment. Normally the earliest year in a cycle is followed by an inter-

Bab King Yr 1=K yr Sun**>	Nbnsr 1 -34.5	Nbnsr 6 -29.8	Nbnsr 11 -25.0	Uknzr 3 -23.0	Mkaid 12 -22.6	Musmk 2 -22.3	Sanh2 3 -18.5	Sssuk 1 -18.2	Sssuk 20 -17.8	Kandl 19 -17.4
Yr 1	-746 U	-741 u	-736 u	-728 u	-709 u	-690 u	-685 U	-666 U!	-647 u	-628 U
2	-745	-740	-735	-727	-708	-689	-684	-665	-646	-627
3	-744 U	-739 a	-734 a	-726 a	-707 A?	-688 a	-683 A'	-664 A	-645 a*	-626 **(?)**
4	-743	-738	-733	-725	-706	-687	-682	-663	-644	**-625 U**
5	-742	-737	-732	-724	-705	-686	-681	-662	-643	-624
6			-731 a	-723 a	-704 A?		-680 a	-661 A	-642 U	-623 A
7			-730 !	-722	-703		-679	-660	-641	-622
8			-729	-721	-702		-678	-659	-640	-621
9				-720 u	-701 u		-677 A	-658 U	-639 U	
10				-719	-700		-676	-657	-638	
11				-718 a	-699 a		-675 a	-656 A	-637 A	
12				-717	-698		-674	-655	-636	
13				-716	-697		-673	-654	-635	
14				-715 a	-696 A		-672 A'	-653 A	-634 A	
15				-714	-695		-671	-652	-633	
16				-713	-694 !		-670	-651	-632	
17				-712 A	-693 a		-669 U	-650 U'	-631 a?	
18				-711	-692		-668 !	-649	-630	
19				-710	-691		-667	-648	-629	
Years	5	5	8	19	19	5	19	19	19	8

FIGURE 5. Neo-Assyrian Intercalations. λ Sun** = tropical longitude at the beginning of Month I of year 1 in that column. A,U = attested XII_2 or VI_2. * = year known, not type. ? = type known, not year. ! = month count from –746 known. Underlined years are the first of new reigns.

calation 2 years later, the exception being in –626, a year of doubtful kingship preceding the accession year of Nabopolassar. Since declaring a year to be intercalary remained the province of kings until the 5th century, it seems likely that the expected intercalation in –626 was put off until –625 when Nabopolassar's reign was firmly established, with the normal pattern resumed in –623. As a result –625 began with the sun –21° from the vernal point, compared with –17° in –628, the only increased in an otherwise diminishing progression since –746. It seems likely that this was an unintended accident due to the change in kingship.

Neo-Babylonian and Early Achaemenid Periods (–620 to –484)

Figure 6 shows intercalations during the Neo-Babylonian and early Achaemenid periods from –620 to –484, after which intercalations were systematically governed by the 19-year cycle. Rather surprisingly, we find much greater irregularity than previously during the half century from year 4 of Nebuchadnezzar (–600) to year 10 of Nabonidus (–545). From –599 to –593 there were 3 successive 2-year intervals between intercalations, while –587 was followed by 4 years between intercalations. –568 was followed by a 5-years interval, which however, was promptly countered by successive intercalations in –563 and –562. Finally after –549 there was again a 4-year interval between intercalations. Clearly something different is happening in this period than either before or (as it turns out) after.

Beginning in –545, year 10 of Nabonidus, regular patterns of intercalation resume, except for the omission of an expected intercalation in –537, the first year of Cyrus's reign at Babylon, which was promptly made up the following year. Again one imagines that the anomaly was a consequence of regime change, especially since normal patterns resume thereafter with cycles of 8 or 19 years, which continued to shrink the distance of the sun from the vernal point at the beginning of each cycle until in –491 it was only –3°.

Bab King Yr 1=K yr> Sun**>	Nbpls 5 -16.4	Nbpls 10 -12.7	Nbpls 18 -10.6	Nbkdr 5 -8.6	Nbkdr 21 -5.6	Nbkdr 40 -5.2	Nbnid 10 -5.8	Cambs 3 -6.4	Dari1 3 -4.4	Dari1 22 -4.0	Dari1 30 -3.0
Yr 1	-620 U	-615 U	-607 U	-599 U	-583 U	-564 **()**	-545 U	-526 U	-518 U	-499 **A**	-491 U
2	-619	-614	-606	-598	-582	-563 **U**	-544	-525	-517	-498	-490
3	-618 A	-613 A	-605 A	-597 U	-581 A	-562 A	-543 A	-524 A	-516 A	-497 A	-489 A
4	-617	-612	-604	-596	-580	-561	-542	-523	-515	-496	-488
5	-616	-611	-603	-595 U	-579	-560	-541	-522	-514	-495	-487
6		-610 U	-602 U	-594	-578 A	-559 A	-540 A	-521 A	-513 A	-494 A	-486 A
7		-609	-601	-593 A	-577	-558	-539	-520	-512	-493	-485
8		-608	-600	-592	-576 A	-557	-538	-519	-511	-492	-484
9				-591	-575	-556 A	**-537** ()		-510 U		
10				-590 A	-574	-555	-536 **U**		-509		
11				-589	-573 U	-554 A	-535 A		-508 A		
12				-588	-572	-553	-534		-507		
13				-587 A	-571 A	-552 A	-533		-506		
14				-586	-570	-551	-532 A		-505 A		
15				-585	-569	-550	-531		-504		
16				-584	-568 A	-549 A	-530		-503		
17					-567	-548	-529 U		-502 U		
18					-566	-547	-528		-501		
19					-565	-546	-527		-500		
Years	5	8	8	16	19	19	19	8	19	8	8

FIGURE 6. Neo-Babylonian and early Achaemenid Intercalations. λ Sun** = tropical longitude at the beginning of Month I of year 1 in that column. Underlined years are the first of new reigns.

Late Achaemenid Period to the Seleucid Era (–483 to –294)

From year 2 of Xerxes' reign or –483 on, intercalations followed a consistent 19-year pattern (Figure 7), which continued to the end of cuneiform records. Since 19-year patterns of intercalation are occasionally evident in earlier periods, we cannot be certain when exactly the practice—which implicitly removed the king from the process—was officially adopted. However, it seems likely to have happened during the reign of Xerxes, or perhaps in the next cycle at the latest. A possible remnant of royal intervention appears in the next three cycles, during the reigns of Artaxerxes I and Darius II, when intercalations of VI_2 months disappeared. Previously and also afterwards, the first year in each cycle was distinguished by an extra VI_2 month, while all other intercalations were XII_2 months. During these three cycles, however, all intercalations were of XII_2 months, perhaps out of concern that Xerxes had unluckily been murdered in a year with an extra VI_2 month. Whatever the reason, the practice of marking the beginning of each cycle with month VI_2 intercalations was resumed under Artaxerxes II, and continued thereafter.

After –483 we have conflicting records for just one intercalation, namely that expected for –383 and attested in two theoretical sources, but also assigned to the previous year in a compilation of Jupiter observations. Since this is the sole anomaly in a century of otherwise consistent practice, it seems likely that the latter account is mistaken, although a plausible explanation of the error is hard to find.

With the uniform adoption of the 19-year cycle to govern intercalations, the progressive slippage of the beginning the first year in each cycle towards the spring equinox effectively came to an end. In –483 the solar longitude at sunset on day 1 of month I was –1.0°, and for the next dozen cycles the solar longitude at the beginning of the first year in each cycle ranged between –1.6° and 0.0°, averaging –0.7°. Since all other years in each cycle began after the equinox, it appears that the object of the exercise was to have all civil years begin on or after the equinox, which would be the case if the equinox were thought to fall 1–2 days earlier than

Bab King	Xerxs	Xerxs	Artx1	Artx1	Dari2	Artx2	Artx2	Artx3	Dari3	S.E.	
Yr 1=K yr>	2	21	19	38	16	16	35	8	4	-1	SS
Sun**>	-1.0	-1.6	-1.2	-0.8	-0.4	0.0	-0.6	-0.2	-0.8	-0.4	(U2)
Yr 1	-483 U	-464 U	-445 **A!**	-426 **A!**	-407 **A!**	-388 U	-369 U	-350 U	-331 U	-312 U	IV 7 U
2	-482	-463	-444	-425	-406	-387	-368	-349	-330	-311	III 18
3	-481 A	-462 A	-443 A	-424 A	-405 A	-386 A	-367 A	-348 A	-329 A	-310 A	III 29 A
4	-480	-461	-442	-423	-404	-385	-366	-347	-328	-309	III 10
5	-479	-460	-441	-422	-403	-384 **(?)**	-365	-346	-327	-308	III 21
6	-478 a	-459 A	-440 A	-421 A	-402 A	-383 **A**	-364 A	-345 A	-326 A	-307 A	IV 2 A
7	-477 (?)	-458	-439	-420	-401	-382	-363	-344	-325	-306	III 13
8	-476	-457	-438	-419	-400	-381	-362	-343	-324	-305	III 24
9	-475 A	-456 a	-437 a	-418 A	-399 A	-380 A	-361 A	-342 A	-323 A	-304 A	IV 5 A
10	-474	-455	-436	-417	-398	-379	-360	-341	-322	-303	III 16
11	-473 A	-454 A	-435 A	-416 A	-397 A	-378 A	-359 A	-340 A	-321 A	-302 A	III 27 A
12	-472	-453	-434	-415	-396	-377	-358	-339	-320	-301	III 8
13	-471	-452	-433	-414	-395	-376	-357	-338	-319	-300	III 19
14	-470 A	-451 A	-432 A	-413 A	-394 A	-375 A	-356 A	-337 A	-318 A	-299 A	III 30 A
15	-469	-450	-431	-412	-393	-374	-355	-336	-317	-298	III 11
16	-468	-449	-430	-411	-392	-373	-354	-335	-316	-297	III 22
17	-467 A	-448 A	-429 A	-410 A	-391 A	-372 A	-353 A	-334 A	-315 A	-296 A	IV 3 A
18	-466	-447	-428	-409	-390	-371	-352	-333	-314	-295	III 14
19	-465	-446	-427	-408	-389	-370	-351	-332	-313	-294	III 25

FIGURE 7. Achaemenid Intercalations. λ Sun** = tropical longitude at the beginning of Month I of year 1 in that column. Underlined years are the first of new reigns.

in fact. This would place the vernal equinox always in month XII, consistent with the Old Babylonian convention, which seems likely to have been the intent. It also suggests that—at least for the purpose of stabilizing the civil calendar—the scribal authorities had a more accurate idea of the date of the spring equinox, placing it roughly 95 days before summer solstice, than is suggested by the so-called Uruk Scheme, which assumed equal seasons of 91 and a fraction days and thus placed the equinox 4 days later.

Figure 8 summarizes the foregoing discussion by plotting the tropical longitude of the sun at the beginning of each year from –746 to –300. Here filled diamonds reflect the effects of

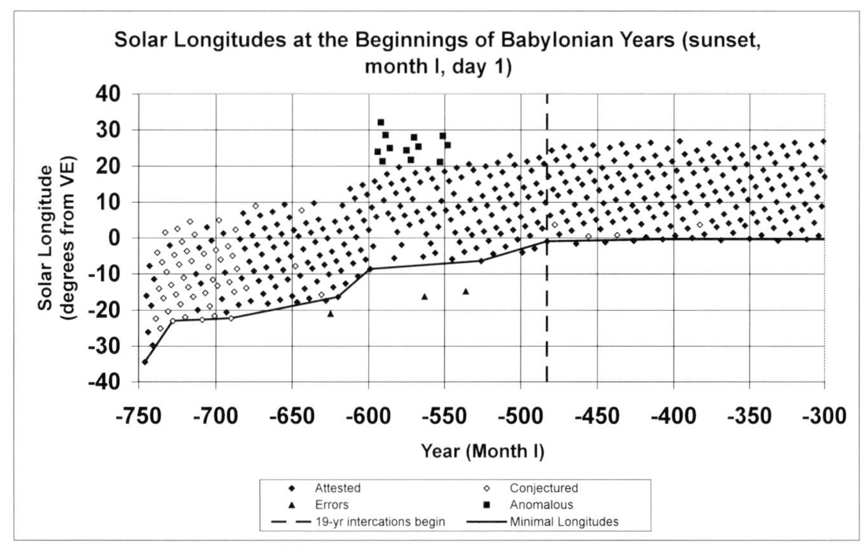

FIGURE 8. Solar longitudes at the beginning (sunset of day 1, month 1) of Babylonian years from –746 to –300.

known intercalations, while open diamonds, largely before –685 are inferred. The triangles far below the normal pattern seem to have been mistakes, which were quickly corrected in subsequent years. The first and last were associated with changes of kings, while the middle one in –563 occurred towards the end of Nebuchadnezzar's reign for no ascertainable reason.

Apart from these errors and the anomalous squares above 20° in the first half of the 6th century (of which more shortly), a striking feature of this chart is its compactness, extending even to the earliest years, and showing that already by the 8th century the intercalation process was controlled and far from random. Also conspicuous is the persistent rise in the band of initial longitudes until the adoption of the 19-year cycle for intercalations in –483. Following a sharp rise from –34° in –746, when the year began too early for even the Assyrian convention, the earliest years' longitudes drift slowly upwards over the next century until 5 Nabopolassar (–620) when they again increase more rapidly, reaching –9° in –599. Thereafter they essentially level off until –526 when a more gradual increase resumes, which effectively ends after –483 with the adoption of the 19-year cycle. The following table gives the details.

Year	$\lambda_0$sun	$\Delta\lambda_0$	ΔY	$\Delta\lambda_0/\Delta Y$
-746 (0 NBNSR)	–34.5°	+12.5	26	+0.49
-720 (11 UKNZR)	–22.0	+ 5.6	100	+0.06
-620 (5 NBPLS)	–16.4	+ 7.8	21	+0.37
-599 (5 NBKDR)	–8.6	+ 2.2	73	+0.03
-526 (3 CAMBS)	–6.4	+ 5.4	43	+0.13
-483 (2 XERXS)	–1.0			

What I believe we see here is a rejection of the Assyrian calendar reform and a gradual program of letting the beginning of the year slide later in the seasonal year with the aim of restoring the ancient calendaric tradition in which the equinox fell in the month named for the barley harvest, i.e. month XII, and the civil year therefore always began on or after the equinox. Departing from this orderly, if uneven, process are the anomalous filled squares between 20° and 30° in the first half of the 6th century, which are accompanied by corresponding gaps in the lower boundary of the chart (Figure 9). These suggest an effort to restore the calendar to the Old Babylonian convention immediately rather than gradually. If so, that effort was quickly countered by a reversion to a more gradual and less discontinuous path of change. This seesawing change recurred three times before the more conservative faction, if that's what it was, won out. In any event it's a curious anomaly in an otherwise gradual and consistent pattern, which occurs at a time when astronomical knowledge and observational practice were advancing sharply.

III. Year Lengths

The steady, apparently purposeful, slide in the initial longitudes (i.e. starting dates in the seasonal calendar) of Babylonian civil years, reflected in Figure 8, implies that the 5- and 8-year intercalation cycles encountered in Figures 6 and 7 were intentional parts of the adjustment process whose inaccuracies were well understood and used to move the yearly band forward more or less rapidly. This brings us to the third part of my subject, namely the year lengths reflected or implied in cuneiform sources and their evolution with time. Basically there are seven empirically distinct values, not counting variants arising from different approximations, which are listed in Figure 10. Quantities placed in brackets are not explicitly attested in the

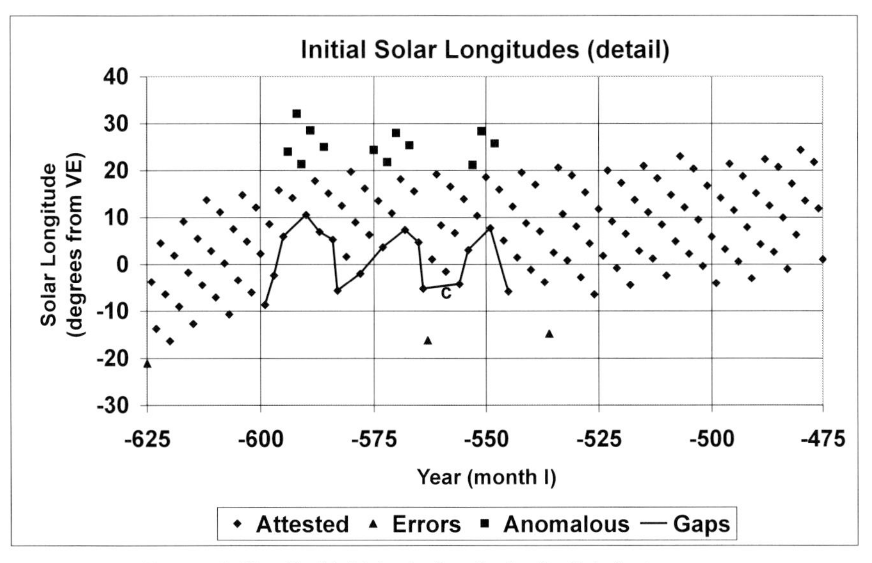

FIGURE 9. Detail of initial solar longitudes for Babylonian years.

	Approx Date	Year (days)	Period Relation	Month	Source
Undifferentiated					
0	<–800	[364]	3y = 37 m	[~29½]	MUL.APIN
1	<–625	364½	[59y = 729 m]	[29½]	BM 36712
2	–600?	365;10	30y = 371m	[30y = 10,955d]	BM 36712
3	–560	365;16	27y = 334m	[30y = 10,958d]	BM 36731
4a	–530	[365;14,50..]	19y = 235m	[29;31,50,18..]	W22805
Sidereal					
5a	–400	356;15,36..	19y-¼° = 235m	29;31,50,6	System A
5b	–300	365;15,33,46	18y+10½° = 223m	29;31,50,8,20	System B
6	–130	365;15,25..		29;31,50,8,20	System B′
Cardinal Phenomena plus Sirius					
4a	–530	[365;14,50..]	19y = 235 m	[29;31,50,18..]	W22805
4b	–350	[365;14,48..]	19y = 235m	[29;31,50,6/8,20]	Uruk Scheme
7	–130	365;14,44,51	297y = 108478d	na	ACT 210
Modern				29;31,50,8,40±20	
		365;14,33..	tropical		
		365;15,23..	sidereal		

FIGURE 10. Empirically distinct year-lengths from cuneiform sources

source cited or otherwise inferred in connection with the derivation of the year-length in question.

The earliest entry on this list, 364 days, derives, as we've seen, from equating 3 years with 37 months, as described in MUL.APIN and assuming 36 months of 29½ days and a 37th month of 30 days. It is more directly suggested by the 8th c. *ziqpu* text (K9794/AO6578), discussed by Horowitz[4] among others,[5] which divides a circuit of the sky into 364 units. I have put it in brackets and labeled it "0", because I do not believe it was ever considered a serious estimate of the year length, since its error accumulates too quickly not to have been noticed. In any event, the record of intercalations after –750 makes clear that already by the 8th century it was well known that the year was longer than this.

The next two year-lengths both appear in the same text (BM 36712), written in Babylon probably early in the 6th century, which is also the earliest example we have of extended arithmetical operations in a text dealing with astronomical subjects.[6] The text, which is broken (as often happens) just where we would most like it not to be, essentially says that an older "measure of the sky" (*mindat šame*) of 364½ days was "lacking". What was lacking turns out to be ⅔ of a day, which leads to 365;10 days as the interval between successive risings of Sirius. The text further says that after 30 years and 371 months, Sirius rises again on the same day of the month.

The measure, 364½ days, is the earliest explicit year-length from cuneiform sources, and might plausibly have been the source of the 364 unit division of the sky in the early ziqpu text. In any event, it clearly predates the 6th century. How seriously it was regarded is hard to know, since it is also a regular number, a fact utilized by the author of BM 36712 in subsequent computations. Its 30-year error of 23 days would suggest either an absence of reliable early records or of close attention. Its most likely origin would seem to be from assuming 1 month equals 29½ days and equating 59 years with 729 months, a period relation also associated with Philolaos the Pythagorean.[7]

With the parameter 365;10 days we arrive at the first solid estimate of the year's length to appear in cuneiform sources. From its context it seems almost certain to have been derived from an estimate that 30 years, marked by Sirius risings, comprised 10,955 days, 20 days longer than implied by the 364½ day year. Equating this interval with 371 months implied a month-length of 29;31,41.. days, which was considerably less accurate than the value (29;31,50,18..days) implied by assuming that 223 months = 6585⅓ days, a relationship employed in Goal Year methods by at least –578. Thus it is possible that the text is earlier than estimated, but also possible and perhaps more likely that no attempt was made to coordinate the two parameters. In any case this parameter seems to date from around the beginning of the 6th century. What is slightly surprising is how relatively late this is, not to mention the fact that the year-length itself is pretty poor.

The next year-length, 365;16 days, appears in BM 36731 another text from Babylon probably written around –560.[8] The text is the earliest ephemeris we know of and consists of computed dates and instants of the solstices and equinoxes together with the dates of the appearances and disappearances of Sirius, probably from the accession year of Nabopolassar (–625) to the end of Nebuchadnezzar's reign in –561. One of several remarkable features of this text is that the solstice and equinox dates are actual dates in the civil calendar, not schematic dates assuming months of 30 tithis, as we typically find in later texts dealing with such phenomena, and indeed even in this text for the risings and settings of Sirius. The basis for this year-length seems likely to have been recognition that Sirius risings recur after 30 years at intervals of 10,958 days, a number 3 days greater than the apparent earlier estimate and correct to the nearest day for sidereal phenomena [accurate = 10957.7d (sid) ..7.4d(trop)].

The Sirius visibilities in this text equate 27 years with 334 months, a period relation first encountered in a text from Babylon (BM 45728), possibly written in 28 Nebuchadnezzar (–576), if not earlier. Combined with the accompanying year-length they imply essentially the same, relatively crude, month-length as that associated with the earlier year length of 365;10 days. Again this is somewhat surprising, since the more accurate month-length implied by the assumed length of the Saros, was certainly known by this date. My guess is that in both cases the period relation was recognized as only approximate, so that no importance was attached to the resulting implicit month-length.

During the second half of the 6th century (if not earlier), it was recognized that equating 19 years with 235 months more accurately described the relationship between months and years than either the 27– or 30–year period relations. The earliest direct evidence of this is a text from Uruk (W22801) which lists schematic dates of summer and winter solstices based on the 19-year cycle from –625 to –530. Remarkably, the summer solstice dates in this text are only ¼ day too early on average for the last 19 years covered by the text—substantially better than the corresponding error in the Babylonian text from 30 years earlier.

How the superiority of the 19-year cycle was established is not known, but any of several considerations might have led to it. First, Sirius risings at intervals of 54 and 60 years would have occurred respectively 2–4 days earlier and later in the month than initially. This noticeable difference and its opposite direction would have implied that 334 months were slightly longer than 27 years and 371 months correspondingly shorter than 30 years, while their combination, equal to 3 19-year cycles would have shown no consistent sensible difference. Even simpler might have been dividing the previously established year length, 365;16 days, by the saros-derived month-length, 29;31,50,18.. days, whose ratio, y = 12;22,8.. m, is closer to that of the 19-year cycle than either of the proximate alternatives. Finally, and simplest of all, would have been noticing that eclipses separated by 235 months recur in the same place in the sky. However arrived at, the superiority of the 19-year cycle was evidently recognized by –530, after which use of the 27– and 30–year cycles effectively disappears. Combining the 19-year period relation with the month-length derived from the saros leads to the 4th year-length in my list, which is marginally less than 365¼ days.

We now skip forward a century and a bit, passing over the adoption of the 19-year cycle to regulate intercalations in the civil calendar early in the 5th century, and the introduction of a uniform zodiac around the middle of that century, to the construction of the first complete mathematical lunar theory known as System A a little after –400. This enterprise required securely establishing the fundamental period relations governing the mean motions of both moon and sun in longitude and thus the year-length. These investigations produced a small adjustment to the 19-year cycle which became the empirical basis for the solar models in both major lunar theories: namely that after 235 months the sun and moon at syzygy fell short of complete returns to their initial (sidereal) positions by ¼°.

This adjustment was almost certainly derived from records of the moon's position near prominent stars during lunar eclipses at intervals of 235, 334, 470 and 804 months. While we know none of the details, plausible estimates with a precision of ½° would have consistently led to the concluded result.[9] Since parallax affects such observations, it is unlikely that this result could have been improved upon without utilizing much longer intervals, for which data was unavailable in –400. In any event it was clearly based on sidereal observations.

This refinement of the 19–year cycle led to the following period relations and related approximations. Thus strictly we have:

$$235 \text{ months} = 19 - 1/1440 \text{ years},$$

whence

$$1 \text{ year} = 12;22,7,56..\text{months}. \qquad (1)$$

In System A theoretical considerations made it desirable that the annual period relation comprise a regular number[10] of years, which led to the natural and convenient approximation,

$$2783 \text{ months} = 225 \text{ years};\ 1 \text{ year} = 12;22,8 \text{ months}. \qquad (2)$$

However expression (2) also implies that

$$223 \text{ months} = 18 \text{ years} + 10;29,10..^\circ \cong 18 \text{ years} + 10;30^\circ,$$

whence

$$1 \text{ year} = 12;22,7,50.. \text{ months},$$

which approximation became the basis for System B's solar model.

System A also incorporated a slightly shorter and more accurate value for the mean month-length of 29;31,50,6 days, which combined with the new empirical period relation leads to the year-length listed as (5a) in my table, 365;15,36..days.

Roughly a century later a second mathematical lunar theory known as System B was developed, comprised almost entirely of schemes and parameters different from those in System A, some of which were improvements. One improvement was the famous value of the mean synodic month, 29;31,50,8,20 days, which was passed to Hipparchus and adopted by Ptolemy. In calculating the mean daily motion of the sun, System B employed a variant of the empirical period relation translated into sarosly terms with the increment in longitude rounded to +10;30°. The resulting daily motion is attested in a procedure text as

$$\delta\lambda_S/d = 0;59,8,9,48,[38.8\cong]40^{\circ/d},$$

which implies that

$$1 \text{ year} = 365;15,33,[45.9\cong]46\text{d},$$

number (5b) in my list. I've labeled them so, to emphasize that both year-lengths derive from essentially the same empirical period relation, to which they apply different, and in System B's case more accurate, month-lengths.

Throughout the time in which both lunar theories were developed and well beyond, solstices and equinoxes continued to be calculated from schemes based on the unadjusted 19–year cycle. While there is no evidence that this reflected more than administrative convenience—much like the calculation of Easter or Passover still today—combining either of the more accurate month-lengths from Systems A or B with the 19-year cycle made the interval between successive solstices or equinoxes 365;14,48.. days as in (4b), which value, truncated to two fractional places, was adopted for the tropical year by Hipparchus and retained by Ptolemy.

The last two year-lengths on my list appear in late texts from Babylon, although two of them reflect orthographical characteristics more typical of Uruk. Number (6) in my list,

365;15,25 days, results from a variant System B solar model (B′), which Neugebauer treats as merely reflecting abbreviated parameters. This variant appears in 4 seemingly late texts, three of which[11] have been tenuously dated to years from –90 to –75. The other, a System B′ ephemeris covering 2 years,[12] cannot be securely dated but also includes instances of orthography from Uruk.

The abbreviated parameters occur in the scheme for monthly solar motion (Column A) whose mean value is $\mu(\Delta\lambda_S)$ = 29;6,20$^{\circ/m}$, which implies the year-length ($360^\circ/\mu(\Delta\lambda_S)\times\mu(m)$ =) 365;15,25[,56].. days. This parameter also appears in a list of year-lengths attributed to Vettius Valens, who in turn ascribes it to "Babylonians" as distinct from Chaldeans. It is very close to Ptolemy's sidereal year-length (365;15,24,31.. days) and consistent with assuming that in 4267 months the sun falls short of 345 revolutions by 6½°, rather than the 7½° attributed by Ptolemy to Hipparchus. In comparison with a tropical year-length based on the 19–year cycle and the System B month-length, it yields values for precession of 36.3′′ per sidereal year and 36.8′′ per tropical year, essentially identical with Ptolemy's result, whose error is due almost entirely to his tropical year length. Thus we may ask whether this was not a new and more accurate parameter rather than simply the result of abbreviation, and if so, whether it was of Greek or Babylonian origin.

The last year-length on my list, 365;14,44,51 days, is from ACT 210, an unusual procedure text from Babylon, which also exhibits orthographical conventions from Uruk. The section in question is labeled "...concerning the revolutions of the moon [returning] to its place ...." and describes the following intervals

[synodic] "months of the moon" [m_B]:
1 m_B = 29;31,50,8,20 days
12 m_B = 354;22, 1,40 days [accurate]
223 m_B = 6585;19,20[,58..] days = "18 years of the moon"

[sidereal] "months of the moon returning to its place" [m_{sid}]
12 m_{sid} = 327;51,20 days [error in last place for 36?]]
241 m_{sid} = 6584;31,20 days [12m_{sid} = 327;51,<u>36</u>..days]
= "18 years of the moon returning to its place"

"18 years of the sun returning to [its ?] after 18 revolutions [$y_?$]
18 $y_?$ = 6584;25,27,18 days
[1 $y_?$ = 365;14,44,51 days, whence
297y = 108,478 days]

The three synodic month intervals all reflect the standard System B month-length, with the length of 223 months, truncated after two fractional places. The stated length of 12 sidereal months leads to an implausibly inaccurate sidereal month-length and is almost certainly in error in its last fractional place which should have been "36" instead of "20". The length of 241 sidereal months is more interesting, since the standard System B parameters make the second fractional place "32" instead of "20", while the System B′ year-length, 365,15,25 days, leads to exactly the result stated. Thus the duration given for 241 sidereal months seems more closely related to System B′ and its nearly correct sidereal year-length, than to the parameters of Standard System B.

In contrast the year-length implied by the interval reported for 18 years of the sun,

365;14,44,51 days and the last in my list, is unrelated to any previously known Babylonian parameters and plays no role in any familiar Babylonian scheme. It is also a conspicuously poor representation of the sidereal year, if that is in fact intended, and at odds with 4 centuries of steady improvement in this parameter. As usual, the critical sign is broken: Neugebauer restores "[*ana ki-š*]*ú*, to its place", which is consistent with the ambient context, but there is scarcely space for these signs, and "[*ana* gu]b, to the solstice", would fit the preserved traces and space as well, if not better. Even more curious is the fact, discovered by Dennis Rawlins,[13] that 297 such years equal 108,478 integral days almost exactly.[14]

This correspondence is otherwise unattested in cuneiform sources, but, as Rawlins notes, it is precisely the interval in whole days between the summer solstice reported by Ptolemy as observed by "the school of Meton and Euktemon" in –431 and that observed by Hipparchus in –134.[15] Ptolemy doesn't report the time of Hipparchus's observation, which should have occurred at noon, 108,478¼ days after Meton's, for consistency with his year length, but Ptolemy's account suggests some inconsistency in the data, and Rawlins has argued that Hipparchus may have observed the solstice a quarter day earlier than consistency requires.[16] An additional curiosity is that according to the Uruk Scheme the two solstices occurred on the dates cited by Ptolemy. Whatever the precise circumstances, it is difficult to avoid the conclusion that knowledge of these observations made its way to Babylon, sometime after –134.

That completes my survey of Mesopotamian year-lengths. Conspicuously absent are either 365 days or 365¼ days, neither of which appear in cuneiform texts to my knowledge. Of the seven described, three clearly made their way into Hellenistic astronomy. First, the standard System B year-length, equivalent to a shortfall of roughly –7½° in 4267 months, was known to and still accepted by Hipparchus after –140. Second, the 19-year cycle combined with the System B month length leads to the value of the tropical year adopted by Hipparchus and accepted by Ptolemy, a correspondence already implicit in a text from Uruk (U107+124)[17] half a century before Hipparchus. Finally, the sidereal year of 365;15,25 days, implicit in System B′ is consistent with the precession and sidereal year adopted by Ptolemy and probably also known to Hipparchus, who appears to have been unable to decide whether this or the more traditional Babylonian year length was more accurate.[18]

In short, interest in the precise length of the year first appears in texts from the 6th century, which reflect rough but progressively better estimates, culminating in recognition that among the short period alternatives the 19–year period relation best describes the relationship between months and years. A century later the development of mathematical lunar theories early in the 4th and 3rd centuries, resulted in improved values for the sidereal year based on the estimate that in 235 months the sun and moon fell ¼° short of complete return. Finally, in texts from the 2nd century or later, we find evidence of an accurate value for the sidereal year, of renewed interest in arithmetical schemes for tropical phenomena based on the unadjusted 19-year cycle, and of input from Greek sources after –134.

Notes

1. This is not quite the same as having the month begin on the evening when the new moon was first visible as is frequently asserted, which given Mesopotamian weather conditions would have led to widely varying month lengths. In fact, no instances of 31 day months are recorded; nor is there any firm evidence of months shorter than 29 days.

2. Englund (1988), pp. 122–133.
3. Huber (1982), pp. 35–6, 57, 59.
4. Horowitz (1998), p. 186.
5. See discussion in Hunger-Pingree (1999), p. 84.
6. Neugebauer and Sachs (1956).
7. Olmstead (1948), p. 210.
8. Originally published in Neugebauer and Sachs (1967) as Atypical Text A. Additional fragments with expanded commentary and corrections are included in Britton (2002), Appendix B.
9. For example an estimated excess of +1° after 334 months (acc = +1;16°), but only +½° after 804 months (acc = +0;55°) would have implied a deficit of –¼° after 235 months (acc = –0;11°), which would have been consistent with an observed deficit of –½° after 470 months (acc = –0;22°).
10. I.e., a number comprised only of the factors 2, 3 and/or 5.
11. ACT 122a (–90); 123(–76/5); and 123aa(–75/4)
12. ACT 123a.
13. Rawlins (1991), pp. 49–51.
14. The multiplication yields 108,478;0,0,27 days.
15. *Almagest* III 1. See Toomer (1984), pp. 138–139.
16. Rawlins (1991), pp. 51ff.
17. See Neugebauer (1947 and 1948) and Britton (2002, Appendix D).
18. According to Ptolemy Hipparchus investigated the period relation, 4267 months = 345 years –7½°. From System A parameters the shortfall is –7;45..°; from System B, –7;20..°; from System B′, –6;30°

References

Britton, J. P., 2002, "Treatments of Annual Phenomena in Cuneiform Sources", in J. M. Steele and A Imhausen (eds.), *Under One Sky: Astronomy and Mathematics in the Ancient Near East* (Münster: Ugarit-Verlag), 21–78.

Cohen, M. E., 1993,*The Cultic Calendars of the Ancient Near East* (Bethesda, MD: CDL Press).

Englund, R. K., 1988, "Administrative Timekeeping in Mesopotamia", *Journal of the Economic and Social History of the Orient* 31, 121–85.

Horowitz, W., 1998, *Mesopotamian Cosmic Geography* (Winona Lake, IN: Eisenbrauns).

Hunger, H., and Pingree, D., 1989, *MULAPIN: An Astronomical Compendium in Cuneiform* (Horn: Berger & Söhne).

——, *Astral Sciences in Mesopotamia* (Leiden, Boston, Köln: Brill).

Olmstead, A. T., 1948, *History of the Persian Empire* (Chicago and London: University of Chicago Press).

Neugebauer, O., 1947, "A Table of Solstices from Uruk", *Journal of Cuneiform Studies* 1, 143–148.

——, 1948, "Solstices and Equinoxes in Babylonian Astronomy during the Seleucid Period", *Journal of Cunieform Studies* 2, 209–222.

——, *A History of Ancient Mathematical Astronomy* (New York, Heidelberg, Berlin: Springer-Verlag).

——, 1955, *Astronomical Cuneiform Texts*, (London: Lund Humphries).

——, and Sachs, A. J., 1967, "Some Atypical Astronomical Cuneiform Texts I", *Journal of Cuneiform Studies* 21, 183–218.

——, and Sachs, A. J., 1969, "Some Atypical Astronomical Cuneiform Texts II", *Journal of Cuneiform Studies* 22, 92–113.

——, and Hunger, H. 1988–1995, *Astronomical Diaries and Related Texts from Babylon*, (Vienna, Österreichischen Akademie der Wissenschaften), Vol. I (-651 to -261), Vol. II (-260 to -164), 1989, Vol. III (-163 to end).

Rawlins, D., 1991, "Hipparchos's Ultimate Solar Orbit and the Babylonian Tropical Year", *DIO* 1.1, 49–66.

Sachs, A. J., 1952, "Sirius Dates in Babylonian Astronomical Texts of the Seleucid Period", *Journal of Cuneiform Studies* 6, 105–114.
——, and Neugebauer, O., 1956, "A Procedure Text Concerning Solar and Lunar Motion: B.M. 36712", *Journal of Cuneiform Studies* 10, 131–136.
Toomer, G. J., 1984, *Ptolemy's Almagest* (London: Duckworth).

The Length of the Month in Mesopotamian Calendars of the First Millennium BC

John M. Steele

Introduction

In the fifth tablet of the Babylonian Epic of Creation *Enūma Eliš*, the god Marduk, having created the universe from the corpse of Tiamat, assigns the sun, moon and stars their places in the heavens. In lines 12–22 Marduk

> made Nannaru (the moon) appear and entrusted to him the night. He designated him as the night's adornment, to define the days. Every month without ceasing he exalted (him) with a crown. "To light up over the land at the beginning of the month, you shine with horns to define six days and on the seventh day a half crown! Be in opposition on the fifteenth day, every half month! When Šamaš (the sun) gazes at you from across the horizon, diminish and recede apace! On the day of disappearance, approach the path of Šamaš! ... on the thirtieth day be in conjunction and rival Šamaš!"[1]

These lines describe the cycle of the moon through the month: the appearance of the lunar crescent at the beginning of the month, half moon on the seventh of the moon, full moon on the fifteenth when the moon and sun are seen on the opposite horizons, and the waning of the moon during the second half of the month until its disappearance and conjunction with the sun on the thirtieth day. Implicit in this description are the calendrical notions of the lunar month whose beginning is determined by whether or not the new moon crescent is seen on the 30th day of the month and the so-called 'ideal' month containing 30 days. These two 'months' may appear to be incompatible since the mean synodic month is a little over 29.5 days, which means that true lunar months contain either 29 or 30 days. The 30-day 'ideal' month, however, has its origins in early administrative practices in which the year containing twelve or sometimes thirteen months each of which may be 29 or 30 days long was replaced by an accounting year of twelve 30-day months to simplify calculation of interest, work obligations and so forth.[2] Brown has plausibly suggested that in *Enūma Eliš* Marduk has created the ideal state of the universe in which lunar months ideally last 30 days, with full moon on the 15th day, and deviations from this ideal situation could be interpreted ominously.[3]

Despite the invention of the 'ideal' calendar, all calendars used for cultic and everyday use throughout Mesopotamia in the second and third millennium BC appear to have used real lunar months. In Sumerian times, Rim-Sin, the ruler of Larsa, described the moon as the god 'Nanna, who establishes the months, who completes the year,'[4] illustrating that the beginning of the month was defined by the appearance of the moon. During the Ur III period there is extensive evidence for monthly festivals taking place on the days of new moon, first half moon and full moon.[5] In some texts of this period the new moon was called 'great crescent at (or of) the head of the month' (u_4-sakar sag-iti-gu-la).[6] In the Old Babylonian period the link between festivals and the lunar month is also attested, for example in the *Atraḫasīs* epic where Enki establishes rituals on the days of new moon, half moon and full moon:

> Enki made his voice heard and spoke to the great gods, 'On the first, seventh, and fifteenth of the month I shall make a purification by washing'.[7]

The association of the first, seventh and fifteenth days with the phases of the moon is referred to also in later works such as the explanatory series i.NAM.giš.ḫur.an.ki.a.[8]

The celestial omen series *Enūma Anu Enlil* also attests to the use of lunar months which begin with the first sighting of the lunar crescent. The first fourteen tablets of the series, which are considered to be a sub-series with its own name IGI.DU_8.A.ME *šá sin* 'the visibilities of the moon', are largely concerned with omens based upon the appearance of the new crescent moon on the first day of the month.[9] The final, fourteenth, tablet of this sub-series contains four tables including one presenting the variable duration of lunar visibility during the month which sets day 1 as the day of first visibility of the moon.[10]

The lunar nature of the calendar is also implied indirectly by literary and artistic evidence. For example, an important hemerology series which sets out the days through the year which are favourable and unfavourable for various activities was named *inbu bēl arḫi* literally 'fruit, lord of the month'.[11] The word *inbu* 'fruit' is used as an epithet of the moon god Sin,[12] so we could translate the title of this series as 'moon, lord of the month'. This association of the moon god with the month again points to the lunar crescent defining the beginning of the month. Indeed, the Akkadian word *arḫu* can mean both 'day of the new moon' and 'month' illustrating once more the juxtaposition of the phenomena of first sighting of the new moon and beginning of the month.[13]

Some medical procedures described in cuneiform sources are to be carried out at specific times during the month.[14] These again attest to the alignment of the month with the lunar phases. For example, a text describing a cure for a migraine says that a ritual must be performed on the fifteenth day, at full moon:

> On the fifteenth day, the day that Sin and Šamaš stand together, you clothe that man in a linen sheet, you incise his temple with an obsidian knife and make his blood flow. You have him sit in a reed hut, you will direct his face to the North. To Sin, to the west, you set up an incense alter of juniper, you libate cow's milk. To the east, you set up an incense altar of cypress, you libate beer. That man will speak as follows: To my left side (is) Sin, the crescent of the great heaven; to my right side (is) Šamaš, the judge, father of the black-headed ...[15]

On the fifteenth day Sin, the moon god, and Šamaš, the sun god, are said to 'stand together'. This expression should be interpreted as meaning that the sun and moon were seen at the same time on the eastern and western horizons respectively, a situation which occurs at dawn on the day of full moon. Thus, the day of full moon is assumed to take place at the middle of the month, on the fifteenth day, implying that the new moon crescent was visible on the first day of the month. In practice full moon will sometimes take place on the fourteenth day of the month, but the text, like the creation epic *Enūma Eliš*, assumes that month will be 30 days long.

It is interesting to note that the moon is referred to as the 'crescent of the great heaven' in the medical text quoted above, despite the fact that the moon is full on the day of the ritual. This designation is frequently used for the moon irrespective of its phase.[16] Iconographically, the moon is most commonly portrayed as a crescent, like a boat, corresponding to the new moon or the period when it is waxing.[17] The prominence of the new moon crescent in these aspects of Mesopotamian culture is very plausibly explained by the importance of the new moon in determining the beginning of the month.

All of the evidence discussed above leads to the conclusion that throughout the Mesopotamian region from at least the third millennium BC the beginning of the month was determined by the sighting of the new crescent moon. However, with the development of astronomical methods in the first millennium that enabled the moon's first visibility to be predicted, the question arises whether the calendar continued to be governed by the observation of the new moon, or whether calculation replaced observation. The use of calculated lunar visibilities would have had tangible benefits such as simplifying the planning of business activities, government administration, even the preparations for cultic activities, but one can easily imagine that a change with tradition may have been unacceptable to Mesopotamian society, which leads to another question: did everyone in Mesopotamia use the same calendar or might there have been parallel calendars used, for example, an astronomical calendar and a cultic calendar? The present paper is intended to address these questions.

The Neo-Assyrian Period

The large body of correspondence sent by Assyrian and Babylonian scholars to the Assyrian kings Esarhaddon and Assurbanipal in Nineveh during the first half of the seventh century BC attest to the significance of establishing the day of the beginning of the month. Many letters and reports were sent to the kings describing the watch for the new moon at the end of the month. Establishing the beginning of the month was of importance for the proper observance of cultic activity. For example, the king and his scholars paid attention to a group of texts known today as 'hemerologies' which set out the favourable and unfavourable days throughout the year to undertake certain activities, as a letter from the chief scribe Issar-šumu-ereš illustrates:

> To the king, my lord: your servant Issar-šumu-ereš. Good health to the king, my lord! May Nabû and Marduk bless the king, my lord! The scribes of the cities of Nin[eveh], Kilizi and Arvela (could) ent[er] the treaty; they have (already) come. (However), those of Assur [have] not (yet) [come]. If it pleases the king, my lord, let the former, who have (already) come, enter the treaty; the citizens of Nineveh and Calah would be free soon and could enter (the treaty) under (the statues of) Bel and Nabû on the 8th day. Alternatively, (if) the king, my lord, orders, let them go, do their work, and get free (again); let them reconvene on the 15th, come here and enter the treaty in the said place at the same time. However, it is written as follows in the hemerologies of the month Nisan (I): "He should not swear on the 15th day, (or else) the god will seize him." (Hence) they should en[ter] the treaty on the 15th day, at d[awn], (but) conclude it only in the night of the 16th day before the stars.[18]

This letter very probably refers to the arrangements for the signing of the oath of allegiance to Esarhaddon following his succession in 672 BC.[19] The hemerological texts cycle through the days of the year giving an indication of whether that day is favourable or unfavourable for activities such as the signing of business contracts, undertaking a legal case, or marriage.[20] Although details are given for 30 days each month, it is clear that these hemerologies do not assume that the 'ideal' calendar is being used; rather they provide the data for the 30th day should it be needed in that month. The use of a true luni-solar calendar is evident from the inclusion of the intercalary months in these hemerologies. By comparing the frequency of business documents and reports of extispicy on the different days of the month with the favourable and unfavourable days given for these activities in the hemerologies, Livingstone has shown that the hemerologies were indeed treated as a guide to practical behaviour in the Neo-Assyrian period.[21] However, in order to use the hemerologies, it was necessary to know which was the first day of the month.

Uncertainty over when the month began could be caused by the occurrence of bad weather. In such cases it seems that the decision over whether to begin the new month on a certain day was made by the king on the advice of his scholars. For example, a letter from Adad-šumu-uṣur, the king's exorcist, informs the king that the moon was too high in the sky on the evening when it was first seen, and advising that the king should check with reports from elsewhere before deciding when the month had began:

> To the king, [my lo]rd: your servant Adad-[šumu-uṣur]. Good health to the king, [my lord]! May Aššur, Sin, Šamaš, [Nabû, Marduk] and the great gods of heaven [and earth] very greatly bless the king, my lord! I observed the (crescent of the) moon on the 30th day, but it was too high to be (the crescent) of the 30th day. Its position was like that of the 2nd day. If it is acceptable to the king, my lord, let the king wait for the report of the Inner City before fixing the date.[22]

There are also several letters and reports sent by the scholars which indicate that the day of first visibility of the lunar crescent and whether the preceding month would be full (30-day) or hollow (29-day) could be predicted in advance.[23] Often these predictions were made in the middle of the month, but a few texts indicate that attempts were made to determine the characteristics of months several months beforehand. For example, a report sent by one of the most prolific Assyrian scholars who wrote many reports concerning the visibility of the new moon crescent, Nabû-aḫḫe-eriba, contains predicted lengths for the following four months:

> [If the moon becomes visible in] Nisan (I) on the 30th day: [Su]bartu [will devour] the Ahlamû; a foreigner will rule the Westland. We are Subartu. If the moon becomes visible on the 30th day: there will be frost in the land. "Frost" is "cold". The moon will be seen with the sun in Tebet (X) on the 14th day. The moon will complete the day in Shebat (XI); on the 14th day it will be seen with the sun. [The moon] will reject the day [in] Adar (XII); [on the xth day] it will be seen with the sun. [The moon] will complete the day in Nisan (I). [From Na]bû-aḫḫe-eriba.[24]

The terminology used in this report to distinguish between full and hollow months is drawn from the practice of watching for the lunar crescent on the evening of the 30th day of the month. The Babylonian day began at sunset so if the new moon was seen an hour or so after the beginning of the 30th day, that day would be 'turned back' or 'rejected' (*turru*) and become the 1st day of the new month. Alternatively, if the moon was not spotted that night then the 30th day was 'completed' or 'confirmed' (*kunnu*) and the new month would start the next evening.[25] There is no evidence that the month was ever longer than 30 days; we must assume that if the moon was not seen at the beginning of the 30th or the 31st day, on account of bad weather, for example, the king would pronounce the new month to have begun anyway.

Despite the evidence for attempts to predict in advance the length of the month, it would appear that the scholars could only advise the king of the likely appearance of the new moon. It remained the king's privilege to accept or reject this advice and proclaim the beginning of the month on the 30th day or the 31st as he chose if the moon had not appeared, just as it was his right to order intercalations.

The Neo-Babylonian and Achaemenid Periods

Most of the evidence for the operation of the calendar in Babylonia during the first millennium BC comes from astronomical texts and dates to after the seventh century BC. In the Neo-Babylonian and Achaemenid periods there are two principal astronomical sources: the 'Astronomical Diaries', which contain night by night reports of astronomical observations and/or predictions, and texts containing compilations of lunar and/or planetary data.[26] The Diaries routinely begin each month with a report of the length of the preceding month and a statement of the length of the time interval between sunset and moonset. This time interval, designated NA, is one of six such intervals: NA at the beginning of the month, ŠÚ, NA, ME, and GE_6 on days around full moon, and KUR at the end of the month on the day when the moon was seen for the last time. These six time intervals are known to modern scholars as the 'lunar six'. Often, following the NA value recorded in the Diary is a statement regarding either the observation of the moon, such as 'it was high/low to the sun' (*ana šamšǐ* NIM/SIG), 'it was bright' (KUR_4), or bad weather which might affect the observation.

The terminology used in recording the length of the month in Babylonian astronomical texts was first deciphered by Epping in 1889.[27] A typical entry begins in one of two ways:

> BAR 1 15 NA
>
> BAR 30 15 NA

BAR is the usual logogram used for *Nisannu* (Month I). This is followed by either the numeral '1' or '30': '1' indicates that the previous month had 30 days; '30' that it had 29. The standard translation of these two cases by Sachs and Hunger is: 'the 1st (of which followed the 30th of the preceding month)' and '(the 1st of which was identical with) the 30th (of the preceding month)'.[28] There follows a value (15 in the examples above) for the time interval NA between sunset and moonset. This time interval is given in units of UŠ where there are 360 UŠ in a day from sunset to sunset. The NA interval may be quoted up to a precision of ⅙th of an UŠ.

The earliest preserved month length recorded in an Astronomical Diary comes from the 37th year of Nebukadnezar II (568–567 BC):

> Year 37 of Nebukadnezar, king of Babylon. Month I, (the 1st of which was identical with) the 30th (of the preceding month), the moon became visible behind the Bull of Heaven; [sunset to moonset: ...][29]

Month	Month Length
I	1
II	30
III	30
IV	1
V	30
VI	1
VII	1
VIII	30
IX	(missing due to damage)
X	(missing due to damage)
XI	1
XII	30

TABLE 1. Month Lengths on BM 33066

Months of 29 and 30 days long are attested from this Diary for months I, III and XI, and months II, XI, XII and month I of the following year respectively. Interestingly, the entry for month I of year 38 indicates that the moon was not seen:

> Year 38 of Nebukadnezar, month I, the 1st (of which followed the 30th of the preceding month): dense clouds so that [I did not see the moon ...][30]

This implies that the month began even though the moon had not been sighted. As this example concerns a full month, it cannot be proven whether the month was predicted to contain 30 days in advance.

Roughly contemporary with this Diary are some compilations of lunar and planetary data containing collections of lunar six data and month lengths of which some were calculated. For example, the tablet BM 33066, frequently referred to as Strass. Camb. 400 after its first publication,[31] contains a complete set of lunar six data dated to year 7 of Cambyses (523–522 BC). Clearly some of these lunar six intervals at least cannot have been observed because of bad weather. The month length data given on the tablet is reproduce in table 1. It is highly likely that some of the month lengths on BM 33066 are also calculated. Furthermore, there is no apparent evidence for a predominance of 30-day months in the winter months, which might be expected if the standard practice was to assume a 30-day month if the moon could not be seen due to bad weather on the evening which begins the 30th day. This text therefore seems to imply that month lengths of both 29 and 30 days could be inserted into the calendar even if the moon could not be seen.

Firm evidence for a 29-day month beginning with a predicted new moon crescent is first found in the Astronomical Diary for 374 BC. The entry for month VIII begins:

> [Month] VIII, (the 1st of which was identical with) the 30th (of the preceding month); sunset to moonset: 10° 30′; mist, I did not see the moon.[32]

Since the moon could not be seen on account of mist the decision to begin month VIII on the 30th evening of Month VII, meaning that month only had 29 days, must have been made on the basis of a prediction of the moon's first visibility. Many further examples of predicted 29- as well as 30-day months are found in later Diaries.

The evidence from astronomical texts proves that in the Neo-Babylonian and Achaemenid periods a luni-solar calendar operated which began with either observed or predicted sightings of the new moon crescent. Was this same calendar used by the general populous, however? Unfortunately, no non-astronomical documents are known which regularly report the length of the month in the same fashion as the Astronomical Diaries. Many economic documents are dated to the 29th and 30th days of particular months, but only those dated to the 30th day provide any information on the length of the month. A document dated to the 29th may have been written on the last day of a 29-day month or the second last day of a 30-day month.

For evidence of 29-day months in the Neo-Babylonian and Achaemenid periods we must turn to a collection of tabular sacrifice records from Uruk dating to the reigns of the sixth century BC kings Nabonidus, Cyrus and Cambyses.[33] These texts, of which forty-one are known, contain monthly ledgers of the number of sheep and goats used in regular sacrificial offerings within the temple. The texts follow a remarkably uniform format consisting of an introductory statement followed by a table divided using vertical rulings, usually into four columns. The first column states the day of the month and in the following columns the number of animals from different sources used as offerings in the daily rituals is recorded. On some dates a comment to the right of the table may provide additional information such as the days on which a special goat sacrifice was made.

The tabular sacrifice records provide direct evidence of the length of the months for which they contain data on the number of sacrificed animals. Of the thirty-nine separate months for which records are preserved (there are two months for which duplicate tablets are known), fifteen contain data for days 1 to 29, twenty-two data for days 1 to 30, and two are damaged at the relevant point.[34] Thus, a little under half of the preserved monthly records are for 29-day months, and the remainder are 30-day months. Theoretically, we would expect the number of 29- and 30-day months to be almost equal over a sufficient length of time; the greater number of 30-day months in this sample is simply due to the small number of preserved texts. Analysis of the tabular sacrifice record by Robbins has also revealed that there was an additional sacrifice at intervals during the month which correspond to the four phases of the moon, mirroring earlier practice.[35]

The practice of cult rituals provides the context for additional evidence of the use of 29- and 30-day lunar months in the Neo-Babylonian and Achaemenid periods. Beaulieu has studied a text which contains an account of the expenditure of sesame oil at the Eanna temple in Uruk for a series of 29-day 'hollow' months (Ì.GIŠ *šá a-na tur-ri* u_4*-mu* a_4 'Sesame oil for the "turning back" of these days').[36] The months given in the text are Month II, IV, V(?), VII, VIII, IX, XI, and XII. Eight 29-day months in a year containing twelve months is more than would be expected, but if we assume that the year contained an intercalary month, the ratio of 29-day to 30-day months is plausible. Beaulieu also publishes and discusses a letter sent by Śameš-ʿidrī of Larsa to an unnamed official at the Eanna temple at Uruk.[37] The letter reports that the officials in Larsa have received the report (presumably from Uruk) that the previous month contained only 29 days and notes that the statue of Šamaš will be clothed on the 15th day of the current month. Beaulieu interprets this text as indicating that the date of the clothing ceremony was pushed back by one day to the 15th if the previous month was hollow, and would normally have taken place on the 14th. However, Robbins has noted that the tabular sacrifice records show no evidence of the adjustment of the dates of cultic activities based upon whether the previous month contained 29 or 30 days.[38]

The letter from Śameš-ʿidrī does, however, indicate that the operation of the calendar in Larsa, and in particular the day of the beginning of the month, was dependent upon communication from the nearby major cultic centre Uruk. A similar situation occurred in other parts of Babylonia. This is demonstrated by a letter sent by one Nabû-šum-iškun, who is presumably writing from somewhere outside the city, to the governor (*šangû*) of Sippar:

> Letter of Nabû-šum-iškun to the *šangu* of Sippar, my father. May Bēl and Nabû decree the well-being of my father. Let me hear quickly a report from my lord as to whether the day is "firm" or "turned back".[39]

The centralisation of the determination of the beginning of the month in the major cult centres probably arose both from practical expediency and from religious and political needs. The celebration of cultic festivals, in particular the *Akītu* or 'New Year' festival, acted as a link between the king, the people and the gods, legitimizing the king's rule.[40] In the first millennium BC the *Akītu* festival began on the first day of *Nisannu* (Month I) and lasted for twelve days. Public cultic activities during the festival included the recitation of the epic of creation, *Enūma Eliš*, and the procession of the king and the sacred statues through Babylon. A centrally organised calendar was needed to ensure that the local population knew which day the public parts of the *Akītu* were to take place. On a practical level, the use of a common calendar no doubt simplified business transactions and the collection of taxes throughout the countryside.

The Seleucid and Parthian Periods

Extensive records of the length of the month during the final three centuries BC are preserved in astronomical texts. Seven main types of astronomical texts containing indications of the length of the month were regularly written during the Seleucid and Parthian periods:[41]

(i) Astronomical Diaries. Night-by-night records of observations of the moon and planets. Where astronomical phenomena were expected but not seen, usually on account of bad weather, a prediction is frequently entered in the Diary instead. Each text normally covers a six or seven month period. All dated Diaries have been published by Sachs and Hunger (1988, 1989, 1996).

(ii) Compilations of lunar and/or planetary data. Collections of observations and predictions of lunar and solar eclipses, planetary phenomena and lunar six data. These tablets may be arranged either as simple lists, or in tablets where each column is separated by a characteristic period (18 years for eclipses, 8 years for Venus, etc.). Most known examples published by Hunger (2001).

(iii) Goal-Year Texts. Collections of planetary and lunar data to be used in making predictions for a coming 'goal-year'. Goal-Year Texts are arranged in sections for each planet and the moon and contain data for each planet or the moon from one characteristic period earlier. All known examples published by Hunger (2006).

(iv) Almanacs. Predicted planetary phenomena (first and last visibilities, stations and acronychal risings), entrances of the planets into zodiacal signs, and lunar and solar eclipses for a given year. Copies of many examples in Sachs (1955).

Month	Diary No. –284	BM 40493
VIII	30	1
IX	1	30
X	1	1
XI	30	1
XII	1	30
I	–	1

TABLE 2. Month Lengths recorded on the Diary No. –284 and BM 40493

(v) Normal Star Almanacs. Predicted planetary passages by certain stars (known as 'Normal Stars'), lunar and solar eclipses, and the lunar six for a given year. Copies of many examples in Sachs (1955).

(vi) Horoscopes. Collection of astronomical data from the time of the birth of an individual. All known examples published by Rochberg (1998)

(vii) Lunar Ephemerides. Lunar data, including the length of the month, calculated using mathematical astronomy. Many examples published in Neugebauer (1955).

In all of these texts the length of the preceding month is designated by a 1 or a 30, meaning the previous month had 30 or 29 days respectively, following a month name. It is informative to compare preserved month lengths in the different groups of texts where they overlap. *A priori* we might expect that the month lengths in the texts containing observations (Astronomical Diaries, lunar and planetary compilations, and Goal Year Texts) would agree but that they will sometimes disagree with those in the texts containing advance predictions (Almanacs and Normal Star Almanacs and the Lunar Ephemerides) if, as has normally been assumed, the beginning of the month was determined by watching for whether the lunar crescent was seen on the evening of the 30th day of the old month.

Before comparing the preserved month length data, however, it is necessary to remark on the unusual recording of data in some Goal-Year Texts. Kugler has shown that in certain Goal Year Texts for years which contain an intercalary month, an intercalary month appears in the individual planetary and lunar sections despite the fact that the years from which the data originated were not intercalary years.[42] Brack-Bernsen has demonstrated that in such cases the lunar six data, which includes the data concerning month lengths, were re-labelled with the name of the following month.[43] She has also shown that the same convention was followed in some of the compilations of lunar six data. For example, in table 2 I compare the month length data for year 27 of the Seleucid Era recorded in the Astronomical Diary No. –284 and the lunar text BM 40493 (= Hunger (2001), No. 37). It is evident that the month lengths have been shifted by one month in BM 40493. The year SE 27 did not contain an intercalary month, but SE 45, eighteen years later, did; it therefore seems plausible to assume that BM 40493 was compiled to assist in the preparation of a Goal-Year Text for SE 45. Brack-Bernsen has shown that the same holds for BM 40091 (= Hunger (2001), No. 40). In the following analysis those texts that contain re-labelled month length data have been adjusted to return the data to its original month.

	Diary	Planetary	Lunar	GYT	Almanac	NSA	Horoscope	Non-Astro
Diary	–	28/0	13/0	35/0	11/0	14/0	2/0	2/0
Planetary	28/0	–	0/0	16/0	5/1	3/0	0/0	1/0
Lunar	13/0	0/0	–	2/0	0/0	2/0	0/0	0/0
GYT	35/0	16/0	2/0	–	0/1	13/0	2/0	1/0
Almanac	11/0	5/1	0/0	0/1	–	13/0	1/0	0/0
NSA	14/0	3/0	2/0	13/0	13/0	–	1/0	0/0
Horoscope	2/0	0/0	0/0	2/0	1/0	1/0	–	0/0
Non-Astro	2/0	1/0	0/0	1/0	0/0	0/0	0/0	–

TABLE 3. A comparison of the month lengths recorded in different types of astronomical texts. For each entry the first value records the number of cases of agreement between month lengths preserved in the different kinds of texts for the same month and the second value the number of disagreements.

Table 3 shows a comparison between preserved month length data in the various Astronomical Diaries, Goal-Year Texts, lunar and planetary compilations, Almanacs, Normal Star Almanacs, Horoscopes, and the very few non-astronomical texts from this period that attest to a 29- or 30-day month. In this analysis I have been able to incorporate a number of month lengths from unpublished Normal Star Almanacs and Almanacs in the British Museum; I am grateful to the Trustees of the British Museum for permission to study unpublished material in their collection. Very occasionally, a Normal Star Almanac gives a second NA value and month length for a certain month when the calculated NA following a 29-day month is close to the visibility limit of the new moon crescent. In these cases, it is clear that the first month length was the accepted value and the second a comment on the visibility situation, so I have taken the former month length in the following analysis. Readings restored by Hunger in his editions of the Diaries have been used without comment since the restoration of these month lengths is determined by the dates of lunar passages by Normal Stars and is therefore secure, except for parts of the Diaries for SE 150 and SE 168 where there is insufficient lunar data to justify his proposed month lengths. For reasons explained the Appendix A I have omitted the lunar text BM 32327+32340 (= Hunger (2001), No. 39) from this analysis.

Comparing first the month length data in the texts which contain mainly observational material (the Diaries, Goal-Year Texts, and lunar and planetary compilations), as expected there are no conflicting month lengths for overlapping texts. Similarly, the texts containing predicted data (the Almanacs and the Normal Star Almanacs), agree where month length data overlaps. More interestingly, however, with only two exceptions, there is agreement between month length data in the 'predicted' texts and the 'observational' texts. The Horoscopes, which were probably based upon data taken from the Almanacs,[44] and the few non-astronomical texts which indicate a 30-day month provide no further conflicts.

Combining the Diaries, lunar and planetary compilations, Goal-Year Texts, and non-astronomical texts in one 'observational' group, and the Almanacs, Normal Star Almanacs, and Horoscopes in another 'predicted' group, we find that there are 48 separate cases of agreement between month lengths in the two groups and only the two exceptions noted above. These two exceptions deserve examining in greater detail. The first conflict is between the planetary compilation BM 45687 (= Hunger (2001), no. 78) and the Almanac BM 33873 published by Kugler (1907), pp. 92ff and Taf. VII. The entry for Month I of SE 130 in the Almanac has '1', implying the previous month contained 30 days, whereas the planetary compilation has '30', implying the previous month contained 29 days. However, the '30' on BM 45687 is placed in half-brackets by Hunger in his edition, meaning that the sign is damaged; thus it is possible that the sign is actually a '1' in agreement with the Almanac. The second disagreement is between the Almanac BM 35551 (= Sachs (1955), no. 1137) and the Goal-Year Text BM 40086+40110 (= Hunger (2006), no. 74). In the Almanac the entry for Month XII_2 is followed by a '1' whereas the Goal-Year Text has '30'. Similar to the preceding case, however, the entry on the Goal-Year Text is placed within half-brackets and followed by a question mark in Hunger's edition, meaning that the reading is very uncertain. It is likely, therefore, that the sign is actually '1' in accordance with the Almanac.

The agreement between the lengths of the months reported in the observational texts and the Almanacs and Normal Star Almanacs leads to the conclusion that during the Seleucid and Parthian periods the day of the beginning of the month was always determined in advance. The method by which this was done is not known for certain, but was probably one of the procedures described on the text TU 11.[45] These methods frequently relied upon the calculation of the lunar six, in particular the interval NA at new moon, through procedures known today as the 'Goal-Year method'. For example, in Section 18 of TU 11 if the value if NA is greater than 12 UŠ then the old month contained 29 days but if it was less than 12 UŠ then it is full; in other words, 12 UŠ was taken to be a visibility limit for the new moon. This rule cannot have been the one used in practice, however, as several examples of calculated NA values less than 12 UŠ are preserved in the Diaries, the smallest being 9;20 UŠ following a 29-day month.[46] However, as there are several other rules on TU 11, it is not surprising that we do not find agreement with this particular rule. Furthermore, it is clear that the month lengths were not determined using the so-called Systems A and B lunar theory found in the ephemerides. Comparision of preserved ephemerides with other texts reveals frequent disagreements between month lengths.

Jones has compared preserved month lengths in the Diaries with those calculated by Parker and Dubberstein using Schoch's visibility theory and found excellent agreement.[47] Before –200 the mean rate of discrepancies is roughly 9% dropping to about 3% after this date. As Jones notes, this suggests a change in the way that the beginning of the month was determined around 200 BC. Discrepancies between month lengths preserved in Babylonian sources and Parker and Dubberstein's tables does not, however, indicate that the start of the month was determined by observation, as argued by Wallenfels,[48] but simply that the Babylonian methods of calculating the first appearance of the lunar crescent was not the same as Schoch's modern theory.

Conclusion

Throughout the last three millennium BC the calendar in Mesopotamia was based upon the lunar cycle with the evening of first visibility of the new lunar crescent marking the beginning of a new month. By the Neo-Assyrian period attempts were being made to predict in advance whether months would be full, containing 30 days, or hollow with only 29 days. In the Neo-Babylonian period we find the first evidence of predicted sightings of the new moon replacing observation in determining the beginning of the month. This practice was fully accepted in the Seleucid and Parthian periods at latest with the character of every month in a coming year predicted in advance.

The adoption of a calendar in which the length of the months were predicted in advance probably came about in large part from practical needs once the ability to make reliable predictions of the day of lunar visibility was achieved. Advance planning for cultic rituals and festivals, which often included the preparation of special foods,[49] would be simplified by the advance knowledge of the day on which the new month began. A counter example illustrates the importance of advance knowledge for preparing for a festival. In the 8th year of Cyrus preparations were made for a kettledrum to be played at the city of Larsa during a lunar eclipse. However, the eclipse did not take place leading to a controversy that is reported in two texts.[50] Brown and Linssen have gone so far as to suggest that improved ability to predict eclipses may have led to an elaboration the related ritual activities.[51]

By moving to a calendar in which month lengths were determined in advance, the problem of transmitting the information that a new month had started from the major cultic centres, perhaps even from the capital of Babylon itself, to regional towns and villages was alleviated. The simplification of administrative practices, both personal and state, therefore also provided a motivation for this change. When this change did take place, it is significant that the source of the calculated month lengths were the schemes based directly upon past observations (or predictions) rather than the mathematical astronomical theories of the ephemerides. One reason for this may have been simplicity: the non-mathematical methods are much easier to use than calculating the numerous columns required in a lunar ephemeris. However, the decision is also in accord with what we know of the predictions of other astronomical phenomena recorded in the Diaries.[52]

Appendix: BM 32327+32340

The tablet BM 32327+32340 (= Hunger (2001), no. 39) is divided into several columns each giving for every month the length of the previous month, the calendar date of the day when the moon set for the first time after sunrise (i.e., the date of NA), and the date of the last visibility of the moon towards the end of the month (i.e., the date of KUR).[53] The tablet also records the dates of the solstices and equinoxes and the first visibility, acronycal rising and last visibility of Sirius, all calculated by the so called 'Uruk Scheme',[54] with the exception that the dates of Sirius's first visibility are all one day earlier than expected.

Comparison of the month lengths given on BM 32327+32340 with those preserved in other sources shows only about 75% agreement (of the 57 overlapping entries, 43 agree and 14 disagree). The disagreements are shown in table 3.

The disagreements between the month length data on BM 32327+32340 come in clumps. This is what would be expected if two different methods were used to calculate the lengths of the months. In the critical cases where the moon's visibility was borderline on the evening of the 30th day of a month, one method, or indeed observation, may make the moon visible and hence the previous month have only 29 days, another make the moon not visible and so the previous month would have 30 days. Over the next few months, however, the two methods will compensate for the borderline visibility and come back into accord. It is therefore likely that the month length data on BM 32327+32340 was calculated using a different method than that used to determine the month lengths used in practice. The occasional differences between the dates of NA and KUR given on BM 32327+32340 and those recorded in the Diaries supports this interpretation.

The use of alternative methods to calculate astronomical phenomena is well known in Babylonian astronomy. For example, the so-called 'Saros Canon' texts of lunar eclipse possi-

Date	BM 32327+32340	Other Sources
SE 69 XII	30	[1]
SE 70 VIII	1	30
SE 70 XII	1	30
SE 71 IX	30	[1]
SE 73 IX	1	30
SE 76 IV	1	30
SE 76 V	30	1
SE 76 VI	1	30
SE 77 VII	30	1
SE 77 VIII	1	30
SE 88 XII_2	1	30
SE 89 VIII	1	30
SE 89 XII	30?	1
SE 93 I	30	1

TABLE 3. Month length data on BM 32327+32340 and in other sources.

bilities reflect a theoretical scheme for determining eclipse possibilities that differs from that used in practice.[55]

Notes

1. *Enūma Eliš* V 12–22; trans. Livingstone (1986), pp. 39–40
2. Englund (1988).
3. Brown (2000), p. 235. See also Robson (2004) and Beaulieu (1993).
4. Cohen (1993), p. 3, citing RIM 4 220.
5. Hallo (1977), Levine and Hallo (1967).
6. Hallo (1977), p. 6.
7. *Atraḫasīs* I 204–207; trans. Dalley (2000), p. 15.
8. K 2164+2195+3510; see Livingstone (1986), pp. 22–23 and 38–42.
9. Verderame (2002a, 2002b). See also Casaburi (2000–01).
10. Al-Rawi and George (1991–92).
11. On the series *inbu bēl arḫi*, see Landsberger (1915), p. 103 and Livingstone (1999).
12. CAD I/J p. 146 sub. *inbu* mng. 1d.
13. CAD A/II pp. 259–261 sub. *arḫu* A mng. 2 and 3.
14. Reiner (1995), pp. 134ff.
15. KAR 184 Rev. 37; trans. Stol (1992), p. 256.
16. Stol (1992).
17. Collon (1992).
18. Parpola (1993), no. 6.
19. Parpola (1983), pp. 3–5. The importance of signing the oath on a favourable day is also stressed in the latter Parpola (1993), no. 5.
20. For examples of Neo-Assyrian hemerologies, see Labat (1939) and Casaburi (2003).
21. Livingstone (1993 and 1997).
22. Parpola (1993), no. 225.

23. Beaulieu (1993), Brown (2000), pp. 198–200.
24. Hunger (1992), no. 60.
25. See Beaulieu (1993) for a discussion of the terminology used to indicate full and hollow months.
26. All dated Astronomical Diaries are published in Sachs and Hunger (1988, 1989, 1996), and the lunar and planetary texts are published in Hunger (2002).
27. Epping (1889), p. 26. See also Weidner (1912).
28. Sachs and Hunger (1988).
29. Diary No. –567 Obv 1.
30. Diary No. –567 Lower edge 1. Although the final few signs are broken away, the restoration is secure.
31. Now published as Hunger (2001), no. 55. For studies of this text, see Kugler (1907), pp. 61–74, Brack-Bernsen (1997), p. 99–103 and (1999), and Britton (2006).
32. Sachs and Hunger (1988), no. -373B.
33. Robbins (1996).
34. Robbins (1996), p. 76.
35. Robbins (1996), pp. 69 and 82.
36. NCBT 1132 edited by Beaulieu (1993), pp. 81–84.
37. NCBT 58 edited by Beaulieu (1993), p. 77.
38. Robbins (1996). p. 82.
39. CT 22 167; Beaulieu (1993), p. 71.
40. Bidmead (2002), Pallis (1926).
41. For the classification of astronomical texts, see Sachs (1947).
42. Kugler (1909–1924), pp. 456–460.
43. Brack-Bernsen (1999), pp. 29–37.
44. Rochberg-Halton (1989).
45. Brack-Bernsen and Hunger (2002); Brack-Bernsen (2002).
46. Diary No. –212B Obv. 1, Diary No. –168A Obv. 16.
47. Jones (2004), pp. 527–528.
48. Wallenfels (1992).
49. Linssen (2004).
50. Beaulieu and Britton (1994).
51. Brown and Linssen (1997).
52. Sachs and Hunger (1988), p. 25.
53. Sachs (1952), pp. 110–112.
54. Neugebauer (1947,1948); Sachs (1952); Slotsky (1993).
55. Steele (2000).

References

Al-Rawi, F. N. H., and George, A. R., 1991–92, "Enūma Anu Enlil XIV and other Early Astronomical Tables", *Archiv für Orientforschung* 38–39, 52–72.

Beaulieu, P.-A., 1993, "The Impact of Month-lengths on the Neo-Babylonian Cultic Calendar", *Zeitschrift für Assyriologie* 83, 66–87.

——, and Britton, J. P., 1994, "Rituals for an Eclipse Possibility in the 8th Year of Cyrus", *Journal of Cuneiform Studies* 46, 73–86.

Bidmead, J., 2002, *The Akītu Festival: Religious Continuity and Royal Legitimation in Mesopotamia* (Piscataway, New Jersey: Gorgias Press).

Brack-Bernsen, L., 1997, *Zur Entstehung der babylonischen Mondtheorie: Beobachtung und theoretische Berechnung von Mondphasen* (Stuttgart: Franz Steiner Verlag).

——, 1999, "Ancient and Modern Utilization of the Lunar Data Recorded on the Babylonian Goal-Year Tablets", in A. LeBeuf and M. S. Ziólkowski (eds.), *Actes de la Vème Conférence de la SEAC* (Warsaw-Gdansk: Insitut d'Archéologie de l'Université de Varsovie), 13–40.

——, 2002, "Predictions of Lunar Phenomena in Babylonian Astronomy", in J. M. Steele and A. Imhausen (eds.), *Under One Sky: Astronomy and Mathematics in the Ancient Near East* (Münster: Ugarit), 5–20.

——, and Hunger, H., 2002, "TU11: A Collection of Rules for the Prediction of Month Lengths", *SCIAMVS* 3, 3–90.

Britton, J. P., 2006, "Remarks on Strassmaier Cambyses 400" (forthcoming).

Brown, D., 2000, *Mesopotamian Planetary Astronomy-Astrology* (Groningen: Styx).

——, and Linssen, M., 1997, "BM 134701 = 1965-10-14,1 and the Hellenistic Period Eclipse Ritual from Uruk", *Revue d'Assyriologie* 91, 147–166.

Casaburi, M. C., 2000–01, "Calendrical Lunar Phenomena and their Interpretation According to Neo-Assyrian Sources: A Preliminary Survey", *Annali Instituto Universitario Orientale* 60–61, 9–57.

——, 2003, *Ūmē ṭābūti "I Giorni Favorevoli"* (Padova: S.a.r.g.o.n. Editrice e Libreria).

Cohen, M. E., 1993, *The Cultic Calendars of the Ancient Near East* (Bethesda, Maryland: CDL Press).

Collon, D., 1992, "The Near Eastern Moon God", in D. J. W. Meijer (ed.), *Natural Phenomena: Their Meaning, Depiction and Description in the Ancient Near East* (Amsterdam: North Holland), 19–38.

Dalley, S., 2000, *Myths from Mesopotamia* (Oxford: Oxford University Press).

Englund, R. K., "Administrative Timekeeping in Ancient Mesopotamia", *Journal of the Economic and Social History of the Orient* 31, 121–185.

Epping, J., 1889, *Astronomisches aus Babylon* (Freiberg im Breisgau: Herder'sche Verlagshandlung).

Hallo, W. W., 1977, "New Moons and Sabbaths: A Case-Study in the Contrastive Approach", *Hebrew Union College Annual* 48, 1–18.

Hunger, H., 1992, *Astrological Reports to Assyrian Kings* (Helsinki: Helsinki University Press).

——, 2001, *Astronomical Diaries and Related Texts from Babylonia. Volume V: Lunar and Planetary Texts* (Vienna: Österriechischen Akademie der Wissenschaften).

——, 2006, *Astronomical Diaries and Related Texts from Babylonia. Volume VI: Goal Year Texts* (Vienna: Österriechischen Akademie der Wissenschaften).

Jones, A., 2004, "A Study of Babylonian Observations of Planets Near Normal Stars", *Archive for History of Exact Science* 58, 475–536.

Kugler, F. X., 1907, *Sternkunde und Sterndienst in Babel* I (Münster in Westfalen: Aschendorffsche Verlagsbuchhandlung).

——, 1909–1924, *Sternkunde und Sterndienst in Babel* II (Münster in Westfalen: Aschendorffsche Verlagsbuchhandlung).

Labat, R., 1939, *Hémérologies et Ménologies d'Assur* (Paris: Libraireie d'Amérique et d'orient adrien-maisonneuve).

Landsberger, B., 1915, *Der Kultische Kalender der Babylonier und Assyrer* (Leipzig: Hinriscs'sche Buchhandlung)

Linssen, M. J. H., 2004, *The Cults of Uruk and Babylon: The Temple Ritual Texts as Evidence for Hellenistic Cult Practice* (Leiden: Brill–Styx).

Levine, B. A., and Hallo, W. W., 1967, "Offerings to the Temple Gates at Ur", *Hebrew Union College Annual* 38, 17–58.

Livingstone, A., 1986, *Mystical and Mythological Explanatory Works on Assyrian and Babylonian Scholars* (Oxford: Clarendon Press).

——, 1993, "The Case of the Hemerologies: Oficial Cult, Learned Formulation and Popular Practive", in Eiko Matsushima (ed.), *Official Cult and Popular Religion in the Ancient Near East* (Heidelberg: Universitätsverlag C. Winter), 97–113.

——, 1997, "New Dimensions in the Study of Assyrian Religion", in S. Parpola and R. M. Whiting (eds.), *Assyria 1995* (Helsinki: The Neo-Assyrian Text Corpus Project), 165–177.

——, 1999, "The Magic of Time" in T. Abusch and K. van der Toorn (eds.), *Mesopotamian Magic: Textual, Historical and Interpretative Perspectives* (Groningen: Styx), 131–137.

Neugebauer, O., 1947, "A Table of Solstices from Uruk", *Journal of Cuneiform Studies* 1, 143–148.

——, 1948, "Solstices and Equinoxes in Babylonian Astronomy", *Journal of Cuneiform Studies* 2, 209–222.

——, 1955, *Astronomical Cuneiform Texts* (London: Lund Humphries).

Pallis, S. A., 1926, *The Babylonian Akîtu Festival* (Copenhagen: Andr. Fred. Host & Son).

Parpola, S., 1983, *Letters from Assyrian Scholars to the Kings Esarhaddon and Assurbanipal: Part II: Commentary and Appendices* (Kevelaer: Butzon & Bercker).

——, 1993, *Letters from Assyrian and Babylonian Scholars* (Helsinki: Helsinki University Press).

Reiner, E., 1995, *Astral Magic in Babylonia* (Philadelphia: American Philosophical Society).

Robbins, E., 1996, "Tabular Sacrifice Records and the Cultic Calendar of Neo-Babylonian Uruk", *Journal of Cuneiform Studies* 48, 61–87.

Robson, E., 2004, "Scholarly Conceptions and Quantifications of Time in Assyria and Babylonia, c. 750–250 BCE", in R. M. Rosen (ed.), *Time and Temporality in the Ancient World* (Philadelphia: University of Pennsylvania Museum of Archaeology and Anthropology), 45–90.

Rochberg-Halton, 1989, "Babylonian Horoscopes and their Sources", *Orientalia* 58, 102–123.

Rochberg, F., 1998, *Babylonian Horoscopes* (Philadelphia: American Philosophical Society).

Sachs, A. J., 1948, "A Classification of the Babylonian Astronomical Tablets of the Seleucid Period", *Journal of Cuneiform Studies* 2, 271–290.

——, 1952, "Sirius Dates in Babylonian Astronomical Texts of the Seleucid Period", *Journal of Cuneiform Studies* 6, 105–114.

——, 1955, *Late Babylonian Astronomical and Related Texts* (Providence: Brown University Press).

——, and Hunger, H., 1988, *Astronomical Diaries and Related Texts from Babylonia. Volume I: Diaries from 652 B.C. to 262 B.C.* (Vienna: Österriechischen Akademie der Wissenschaften).

——, and Hunger, H., 1989, *Astronomical Diaries and Related Texts from Babylonia. Volume II: Diaries from 261 B.C. to 165 B.C.* (Vienna: Österriechischen Akademie der Wissenschaften).

——, and Hunger, H., 1996, *Astronomical Diaries and Related Texts from Babylonia. Volume III: Diaries from 164 B.C. to 61 B.C.* (Vienna: Österriechischen Akademie der Wissenschaften).

Slotsky, A. L., 1993, "The Uruk Solstice Scheme Revisited", in H. D. Galter (ed.), *Die Rolle der Astronomie in den Kulturen Mesopotamiens* (Graz), 359–365.

Steele, J. M., 2000, "Eclipse Prediction in Mesopotamia", *Archive for History of Exact Science* 54, 421–454.

Stol, M., "The Moon as Seen by the Babylonians", in D. J. W. Meijer (ed.), *Natural Phenomena: Their Meaning, Depiction and Description in the Ancient Near East* (Amsterdam: North Holland), 245–278.

Verderame, L., 2002a, "Enūma Anu Enlil Tablets 1–13", in J. M. Steele and A. Imhausen (eds.), *Under One Sky: Astronomy and Mathematics in the Ancient Near East* (Münster: Ugarit-Verlag), 447–457.

——, 2002b, *Le Tavole I–VI della serie astrologica Enūma Anu Enlil* (Rome: Grafica Cristal).

Wallenfels, R, 1992, "30 Ajjaru 219 SE = 19 June 93 BCE", *NABU* 1992–46.

Weidner, E., 1912, "Der dreissigste Tag", *Zeitschrift für Assyriologie* 27, 385–388.

On Greek Stellar and Zodiacal Date-Reckoning

Alexander Jones

As is well known, the calendars employed in Greek communities for civil and cult purposes before the Roman period were lunisolar, and subject to much variation from place to place in the naming of months and days and in the season in which the year's beginning was approximately fixed. The evidence for how these calendars were administered is extremely uneven in both quantity and quality.[1] Even in the case of the Athenian calendar, for which we have by far the most extensive documentation, consensus still cannot be reached on whether—and if so, when—a fixed cycle of intercalations was adopted, nor can attested dates be synchronized with our Julian chronography with anything approaching the precision (generally plus or minus one day at most) possible for contemporary Babylonian and Egyptian dates.[2] Lunisolar intercalation cycles are associated in ancient Greco-Roman texts with various astronomers, including Meton (late 5th century BC, 19 years = 235 months), Eudoxus (early 4th century BC, 8 years = 99 months), Callippus (late 4th century BC, 76 years = 940 months = 27759 days), and Hipparchus (late 2nd century BC, 304 years = 3760 months = 111035 days).

Of these cycles, the Callippic 76-year cycle is the only one around which we can confidently say that a full-blown and widely used calendar was constructed, possibly by Callippus himself.[3] This Callippic calendar employed the month names and midsummer epoch of the Athenian calendar, but the years are designated simply by year number within a serially numbered 76-year period (e.g. "according to Callippus period 3, year 14") without reference to the conventions for naming years in any specific locality, such as names of magistrates or regnal years. All the attestations of Callippic dates, which range from the early 3rd century BC to the middle of the 1st century AD, are in the context of astronomical observation reports or predicted astronomical data, so that it is legitimate to speak of the Callippic calendar not merely as an astronomical calendar but as an *astronomers'* calendar, which nevertheless retained and regularized the key features of the Greek local calendars on which laymen (i.e. nonastronomers) depended.

The Egyptian calendar, in its pre-Roman unintercalated form, can also be regarded as an astronomers' calendar when it appears in astronomical and astrological texts and tables, though in this instance we are not dealing with a calendar structure expressly devised for astronomical purposes on the model of lay calendars, but rather an actual civil calendar that astronomers, apparently as early as Timocharis in the 3rd century BC., found convenient, and that thus survived in this specialized context long after laymen had abandoned it. In one important respect, however, the astronomers' version of the Egyptian calendar was not simply taken over from common usage. Hipparchus, who unlike his predecessor Timocharis and his successor Ptolemy did not work in Egypt, numbered specific Egyptian calendar years not by Ptolemaic reigns but by an era count from the death of Alexander, while Ptolemy (perhaps following a Hipparchian precedent) adapted a Babylonian king list for the designation of Egyptian years for the interval from the mid 8th century BC to the death of Alexander. These are conventions that would have had no application outside astronomy.

One last astronomers' calendar is known from a set of observation reports in Ptolemy's *Almagest*. This calendar "according to Dionysius" had, so far as we know, a shorter and more localized career than the Callippic and the astronomical Egyptian calendars, since all the attestations fall within a few decades during the 3rd century BC and appear to originate in a small group of astronomers probably working in Egypt. The Dionysian calendar stands out among the Greek astronomers' calendars in that it appears not to be an adaptation of an already existing kind of calendar: its years are solar, and subdivided into months correlated with the sun's passage through the zodiacal signs, with no reference to the moon. Nevertheless the Dionysian calendar has antecedents in Greek practices of time reckoning, most immediately in a technical astronomical application but ultimately in broader contexts of intellectual life – but these antecedents were not calendars. The present paper is an attempt to sketch the relations and development of these practices, in which the determining elements are the sun, stars, and later on the zodiac.

1. Lay Dating by Annual Astronomical Phenomena

After the Pleiades, there was good weather with clouds and mists. [Sick people had] crises [i.e. critical stages of illness] on the fifth day and sixth day and seventh day and also at greater intervals. The fevers were characterized by relapses, wandering about to some extent, loss of appetite, and bile. And there were dysenteries accompanied by loss of appetite and fever. Around the setting of the Pleiades, the southerly winds blew strongly. There were hemorrhages and tertian fevers and agues. The fellow in the cobbler's shop had a hemorrhage; a little violent excretion; he reached crisis on the seventh day with shivering. The boy living by the last shop had an abundant hemorrhage on the fourth day; at once he began to babble; the belly was constipated; the abdomen was painful and hard; a suppository on the sixth day brought nasty yellow stuff; early on the seventh day, tossing about, much shouting, pulsation of the blood vessels by the navel...

At the winter solstice, a meteor, and not a small one; on the fifth and sixth days after, earthquake. When we got to Perinthos, [we saw] the asthmatic woman, the wife of Antigenes, who did not know if she was pregnant, manifesting red discharges from time to time; small belly, at other times big, for example when she went for too fast a walk (for she had a cough). It was her eighth month. She settled down after first having a fever. [pseudo-Hippocrates, *Epidemics* 4.20]

The majority of fishes breed during the three months Mounychiôn, Thargêliôn, and Skirrophoriôn. A few do so in the autumn; for example the *salpê*, the *sargos*, and others of that sort [breed] a little before the autumnal equinox, and the *narkê* and *rhinê*. A few breed in either winter or summer, as was said earlier; for example the *labrax*, the *kestreus* in winter, the *belonê* in summer around [the month] Hekatombaiôn, and the *thynnis* around the summer solstice. [Aristotle, *Historia Animalium* 5.11]

Before [the time of] Antigenides, when they used to play the *aulos* [i.e. shawm] in an unsophisticated manner, [people say that] the season for cutting [the reeds] was at [the rising of] Arcturus in the month Boêdromiôn; [a reed] cut thus becomes usable quite a few years afterwards and needs a lot of preliminary blowing, but the gap between the tongues is closed up, which is useful for [*text corrupt*]. But when they went over to the sophisticated manner, the [practice of] cutting was also changed. They now cut it in Skirrophoriôn and Hekatombaiôn, as it were a little before the [summer] solstice or at the solstice; they say that it is usable when three years old and needs [only] brief blowing, and that the tongues withstand rough vibrations, which is necessary when people play the *aulos* in the sophisticated manner. [Theophrastus, *Historia Plantarum* 4.11.3]

Histories of Greek astronomy often take Hesiod's didactic poem, *Works and Days* (composed around 700 BC) as their effective starting point.[4] Hesiod delineates an agricultural year in terms of a cycle of natural "signal" events, many of which (though not all) are astronomical: the morning risings and settings of Orion and the Pleiades, the morning and evening risings of Arcturus, the morning rising of Sirius, and the two solstices. He also states intervals in days for the times from the evening setting to the morning rising of the Pleiades (40 days) and from the winter solstice to the evening rising of Arcturus (60 days) but not a complete scheme connecting all the dates of the phenomena; proper times are fundamentally determined by observations, not by counting days. Just once Hesiod mentions a lunar calendar month, Lênaiôn, a time of harsh wintry weather when activities are to be avoided. (Interestingly, Lênaiôn is not a month of the Boeotian calendar, which would have been local to Hesiod and his immediate audience, but Ionian.) Thus he presumes that a lunar calendar will receive intercalary months with sufficient regularity to keep a particular month around a particular stage of the natural year; but in general he seems intent on not relating the agricultural year to a lunar calendar.

It is less commonly recognized that this method of astronomical dating independent of lunar calendars was practiced also in the classical Greece of the fifth and fourth centuries BC. The passages quoted at the beginning of this section are illustrations of the two chief contexts in which such datings occur. The first is a specimen of the case notes recorded by an itinerant Greek physician who probably lived in the late fifth or early fourth century BC, and preserved as Book 4 of the *Epidemics* in the Hippocratic corpus. The seven books of *Epidemics*, which appear to derive from the work of several such physicians working in many places in the Greek world about this period, contain numerous statements of the times of year when certain patterns of weather were observed and individuals and groups of people fell sick.[5] Although these were records of specific historical occurrences, not generalized patterns, the authors eschew any identification of the year or other absolute chronological data, and they never employ the local calendars of the localities where they were working. Instead, they either refer simply to the meteorological season (*eär* = spring, *theros* = summer, *phthinopôron* = autumn, *kheimôn* = winter) or, more precisely, to one of a cycle of seven annual astronomical events:

winter solstice
vernal equinox
summer solstice
morning rising of Sirius (about 30 days after summer solstice)
morning rising of Arcturus (about 80 days after summer solstice)
autumnal equinox
morning setting of Pleiades (about 50 days after autumnal equinox)

Nowhere in the *Epidemics* do we find an explanation of why the medical histories were dated with reference to astronomical phenomena, and to these phenomena in particular. Galen (2nd century AD) was confident that he knew the reason:

> If all nations had the same [months], Hippocrates would not have referred to Arcturus and the Pleiades and Sirius and the equinoxes and solstices, but he would have been content to state that such and such conditions occurred in the makeup of the environment at the beginning of [the month] Dios, naming [the month]

> according to the Macedonians if that was the state of affairs [i.e. if all nations used Macedonian months]. But since in fact a reference to Dios is clear only to the Macedonians, but not to the Athenians and the rest of mankind, whereas Hippocrates intended to be of service to people from all nations, it was better for him to record just the equinox without mentioning in which month. For the equinox is a universal matter, while the months are local to each nation. Now anyone who is ignorant of astronomy should really know that he is not obeying Hippocrates' exhortation to [learn] this [science] for the sake of using the aforesaid [phenomena]. [Galen, *In Hippocratis librum primum epidemiarum commentarii iii*, 19]

In other words the only thing that prevented Hippocrates—for Galen believed that the historical Hippocrates was the author of Books 1 and 3 of the *Epidemics*, while at least three of the others were by his son Thessalus—from dating everything according to a conventional civil calendar was that no single calendar would have been familiar to all of his intended readership.

A little further down in the same commentary, Galen observes rather casually that it is only possible to assign fixed dates for the solstices, equinoxes, and stellar phenomena in a calendar based on solar years:

> Of course the months must not be counted according to the moon as they are in most of the Greek cities at present and in all of them in olden days, but according to the sun as they are counted among many nations and in particular among the Romans... [Galen, *In Hippocratis librum primum epidemiarum commentarii iii*, 21]

Hence if "Hippocrates" had dated the events recorded in the *Epidemics* with reference to the months of, say, the calendar of Cos, even a reader from Cos would not have been getting exactly the same information as is conveyed by the astronomical datings, since in even the most uniformly intercalated lunar calendar an annual astronomical phenomenon can fall anywhere within an interval of approximately thirty calendar days in different years. This might or might not be a strong motive for avoiding civil calendars, depending on whether the authors of the *Epidemics* meant the correlation of medical phenomena with recurring stages of the natural year to be rough or precise.

In another roughly contemporary text in the Hippocratic corpus, *Airs, Waters, Places*, we are warned to avoid performing therapeutic actions close to changes of the seasons:

> Especially one ought to watch out for the most important changes of the seasons and neither administer a drug, if one has the choice, nor apply any cautery or cutting to the bowels before ten days have passed or even more. The greatest and most dangerous are these four: both the solstices and especially the summer solstice, and both the [dates] regarded as equinoxes and especially the autumnal equinox. One ought also to watch out for the risings of the stars and above all that of Sirius, and thereafter that of Arcturus, and moreover the setting of the Pleiades. For illnesses have their crises especially on these days; some cause death, others cease, and all the rest change into another form and another condition. [pseudo-Hippocrates, *Airs, Waters, Places*, 11]

The list of medically significant astronomical events exactly corresponds to the ones cited in the *Epidemics*. The last quoted sentence is surely the key to understanding the astronomical dates in the *Epidemics*: they are not a substitute for the civil calendars, but an expression of a theory tying together astral phenomena, the meteorological environment, and patterns of illness.

In still another Hippocratic text, *Regimen* Book 3, the appropriate diet and routine for maintaining good health through the course of the year is indexed according to the following cycle of astronomical phenomena:

	Morning setting of Pleiades = beginning of winter
44 days	Winter solstice
59 days	Evening rising of Arcturus
32 days	Vernal equinox = beginning of spring
48 days	Morning rising of Pleiades = beginning of summer
[*lost interval*]	Summer solstice
93 days	Morning rising of Arcturus = autumnal equinox = beginning of autumn
48 days	Morning setting of Pleiades

Though similar to the cycle of phenomena in the *Epidemics*, this cycle omits Sirius but has two (out of a possible four) phenomena for both Arcturus and the Pleiades. The other noteworthy element here is the more or less complete scheme of time intervals between the phenomena, which can be seen as a step towards eliminating the need for observing the phenomena.

Neither the Hesiodean farmer nor the Hippocratic physician could by any reasonable definition be called an astronomer, but their time-reckoning practices presume some astronomical knowledge. One should not exaggerate the degree of this knowledge, however. For the stellar phenomena one would have to be able to pick out three or four very bright or easily recognized stars or star groups—admittedly close to sunrise or sunset, when most of the background of dimmer stars would be invisible. Solstices present no difficulties so long as one is not attempting to determine the exact day, and there is nothing in these texts to suggest anything more sophisticated than watching for when the sun's rising or setting point on the horizon appears to remain unchanged for several successive days. The equinoxes are not quite as trivial to observe even crudely, so it is interesting to find them routinely cited in the *Epidemics*. The author of *Airs, Waters, Places* acknowledges that the physician must to some extent take these dates on trust when he refers to them as the dates "regarded as" (*nomizomenai*) equinoxes. I imagine that this means that one did not attempt to observe the equinoxes directly by, e.g., watching the sun's rising or setting points, but instead one counted a certain number of days from the preceding solstice or from one of the stellar dates.

When Aristotle and Theophrastus wanted to indicate the times in the natural year pertaining to the life cycles of animals and planets (and to human activities connected with them), their practice was less consistent than that of the Hippocratic writers. Instead, they sometimes refer to months of the Athenian calendar (e.g. Hekatombaiôn, Boêdromiôn, Mounychiôn, Thargêliôn, and Skirrophoriôn in the passages quoted), sometimes to solar and stellar phenomena, and often, redundantly, to both. The cited phenomena include all the solstices and equinoxes, the morning rising of Sirius, the morning rising and setting of the Pleiades, and the morning rising and evening setting of Arcturus. Both authors seem to have expected their readers to be familiar with the approximate correlation of the astronomical events and the calendar months. In general it is difficult to see the reason why one mode of dating or the other is being used. On one occasion (*Historia Animalium* 5.22) Aristotle asserts that bees deposit honey primarily at the dates of risings of constellations, but in general neither Aristotle nor Theophrastus suggests that the astronomical phenomena are more than convenient ways of marking stages of the year.

While the Hippocratics and the Peripatetics shared a concern with naturally occurring phenomena that suggests an obvious motivation for astronomical dating, one might not expect to find this kind of dating in historiographical writing; yet there are a small number of instances in Thucydides: two references to events taking place "around" the winter solstice (7.16 and 8.39), and one to an event "around" the morning rising of Arcturus (2.78). Thucydides generally organizes time by broader references to the seasons, and in fact employs Athenian calendar months every bit as rarely as the astronomical markers (2.15, 4.118, 5.19).

The high degree of consistency in the choice of phenomena exploited for "natural" dating by Hesiod, the Hippocratic authors, the Peripatetics, and Thucydides deserves comment. One finds always the same few stellar objects (Sirius, Arcturus, the Pleiades, Orion only in Hesiod), the solstices, and—except in Hesiod—the equinoxes. A further indication that one was dealing with an extremely restricted list of familiar events is the elliptical manner in which they are often specified: it was enough to say "after Arcturus" (*met' Arktouron*) if one intended the days following the star's morning rising.

2. The Parapegmatists

> And I have recorded the episêmasiai belonging to these [stars] and set them down according to the Egyptians and Dositheus, Philippus, Callippus, Euctemon, Meton, Conon, Metrodorus, Eudoxus, Caesar, Democritus, [and] Hipparchus. Of these, the Egyptians observed here, Dositheus in Cos, Philippus in the Peloponnese and Locris and Phocis, Callippus on the Hellespont, Meton and Euctemon at Athens and the Cyclades and Macedonia and Thrace, Conon and Metrodorus in Italy and Sicily, Eudoxus in Asia and Sicily and Italy, Caesar in Italy, Hipparchus in Bithynia, Democritus in Macedonia and Thrace. [Ptolemy, Phaseis ed. Heiberg 66–67.]

> Why does the *notos* wind blow at the [rising of] Sirius, and this occurs as regularly as anything else? Is it because the region below is hot because the sun is <not> far away, so that the vapour is abundant? They would actually blow often were it not for the Etesian winds; but as things are, [the Etesians] prevent them. Or is it because a sign occurs [*sêmainei*] at all the settings and risings of the stars, and not least at this one? Obviously there are winds especially at this [the rising of Sirius] and after it. When it is stifling, winds, and indeed the hottest ones, are set in motion at this [rising]; and the *notos* wind is hot. But since there is a tendency especially for changes to occur from opposites to opposites, and the *prodromoi* winds, which are *boreai* winds, blow before [the rising of] Sirius, it makes sense that *notos* blows after [the rising of] Sirius, since an *episêmasia* occurs [*episêmainei*], and for stars making their rising "an *episêmasia* occurs" means causing a change in the air; and all winds change into the winds that are opposed or to their right. [pseudo-Aristotle, *Problemata* 26.12, 941b.]

The restricted choice of astronomical phenomena used by the lay writers is the more remarkable because by the time when Aristotle and Theophrastus wrote (the third quarter of the fourth century), a tradition of astrometeorology making use of a considerably larger repertoire of stars and constellations had been in existence for about a hundred years if not longer.[6] Our evidence for this tradition is from Hellenistic and Roman period documents that are now conventionally referred to as *parapegmata*, in which solar and stellar phenomena as well as presumed annual repetitions of weather changes are assigned to specific days within an idealized solar year. Two of these documents are especially important because they contain numerous dates of phenomena attributed to past authorities: the so-called "Geminus" parapegma that is preserved immediately following the abrupt (and perhaps mutilated) end of Geminus' *Isagoge*, and Ptolemy's *Phaseis*. The "Geminus" parapegma, which probably has no connection with Geminus himself, cites no authority later than the third century BC, whereas Ptolemy used sources as late as Julius Caesar. Unfortunately Ptolemy chose to include only the weather dates from his authorities, preferring his own calculations for the astronomical phenomena.

The earliest persons associated with parapegmatic data are Meton, Euctemon, and the Presocratic atomist Democritus, all of whom were active in the late fifth century BC; the authenticity of the reports associated with Democritus is, however, highly questionable, and anyway most of them pertain to weather and only a few to astronomical phenomena. Several of the ancient references to Meton and Euctemon link their names as if they were colleagues, but the nature of their collaboration—if that is what it was—is unclear, and the parapegmatic data are always attributed to just one man or the other. As it happens, only a single astronomical event, the morning rising of Sirius, has an extant parapegmatic entry ascribed to Meton.

For Euctemon, however, the "Geminus" parapegma preserves for us what seems to be a large fraction of the complete set of astronomical phenomena that he recorded.[7] In addition to the solstices and equinoxes, Euctemon's list of phenomena included morning or evening risings or settings of fifteen bright stars and constellations; in many instances "Geminus" has all four possible phenomena, and it seems likely that this was originally true for more of them before some entries dropped out in the transmission (either between Euctemon and "Geminus" or in the manuscript tradition of the "Geminus" parapegma itself). Table 1 is a list of Euctemon's stars and constellations, roughly in order of decreasing brightness. For the constellations we identify the brightest star or stars, since these were probably the ones whose appearances or disappearances determined the dates of risings and settings for the constellations as a whole.

It is a curious list, containing almost all the stars visible at the latitude of, say, Athens down to magnitude 1 (Procyon and Spica are the exceptions), omitting practically all stars with magnitudes between 1 and 2 (e.g. Pollux, Fomalhaut, Deneb, Regulus, Adhara, Castor, to name only the brightest), but then including several distinctly less bright groups. Some of these dimmer groups are fairly compact and hence recognizable asterisms, but Pegasus covers such a wide area as to make it doubtful which stars Euctemon had in mind. Euctemon's significant stars and constellations are fairly well spread in right ascension, and too scattered in declination to constitute a continuous belt. There does not seem to be any close relation between his choice of stars and any of the known Mesopotamian lists.

The "Geminus" parapegma also attributes to Euctemon many statements of weather changes (for which the technical term was *episêmasiai*, "significations"). Like all the data in this parapegma, these events are dated according to a rigid count of days that begins with the summer solstice. A second source for weather changes, but not astronomical phenomena, attributed to Euctemon is Ptolemy's *Phaseis*, in which all events are dated according to the Alexandrian (reformed Egyptian) calendar, which had a four-year cycle of 365-day and 366-day years like the Julian calendar and thus remains approximately fixed in relation to the solar year.

The relationships among these three sets of dates is easiest to see if we plot them along a scale of days counted from the summer solstice as day 1 (figure 1). First of all, nearly every

Star/Constellation	Brightest stars	Magnitude
Sirius	(α CMa)	–1.46
Arcturus	(α Boo)	–0.04
Lyra	Vega (α Lyr)	0.03
Capella	(α Aur)	0.08
Orion	Rigel (β Ori)	0.12
	Betelgeuse (α Ori)	0.50
Aquila	Altair (α Aqu)	0.77
Hyades	Aldebaran (α Tau)	0.85*
Scorpius	Antares (α Sco)	0.96
	Shaula (λ Sco)	1.63
Corona Borealis	Aphekka (α CrB)	2.23
Pegasus	Enif (ε Peg)	2.39
	Scheat (β Peg)	2.42
	Markab (α Peg)	2.49
	Algenib (γ Peg)	2.83
Vindemiatrix	(ε Vir)	2.83
Pleiades	Alcyone (η Tau)	2.87
Sagitta	(γ Sge)	3.47
Delphinus	Rotanev (β Del)	3.63
Haedi	(ζ Aur)	3.75

* According to later conventions Aldebaran was not counted among the Hyades, but the visibility dates assigned to the Hyades in the parapegmata seem to require the inclusion of this bright star.

TABLE 1. Euctemon's stars and constellations

date for which there is an astronomical event also has a weather change in the "Geminus" parapegma, and *vice versa*. This implies that for Euctemon, the astronomical events were not just a tool for tracking the progress of the natural year, as Geminus claimed in his factitious account of how the parapegmatic tradition began (quoted at the head of the next section below), but immediate signs, possibly even causes, of the weather changes.[8]

Secondly, while it is not possible to line up Ptolemy's Alexandrian calendar dates with the "Geminus" parapegma's day count from summer solstice in such a way as to make the two sets of weather phenomena exactly match, we can get nearly two thirds of the parapegma's dates to match dates from Ptolemy—21 coincidences out of 35 dates in "Geminus" and 55 in Ptolemy—if we equate Euctemon's summer solstice with Ptolemy's Epeiph 2 (equivalent to June 26 in any Julian calendar year). In most of these instances of coinciding dates the specifics of the weather changes are substantially the same in both sources. This is certainly the alignment that Ptolemy made, but it is not clear why he did it in just this way. In agreement with his solar theory, he set the summer solstice for his own period, the mid second century A.D., as occurring on Epeiph 1 (June 25). According to *Almagest* 3.1 the length of

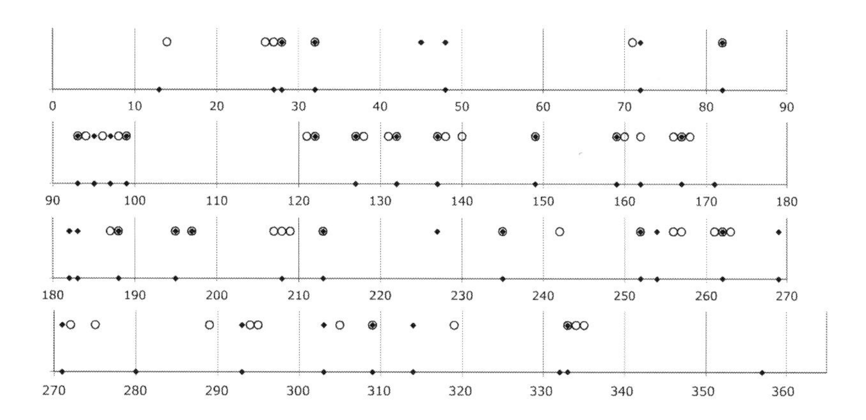

FIGURE 1. Concordance of parapegmatic dates attributed to Euctemon according to dates counted from summer solstice. Lower line: astronomical phenomena in the "Geminus" parapegma. Upper line: weather phenomena in the "Geminus" parapegma (solid) and in Ptolemy's Phaseis (circles).

the tropical year is $365\frac{1}{4} - \frac{1}{300}$ days. Hence since Euctemon worked nearly six hundred years before Ptolemy, his solstice should have fallen on Epeiph 3, not Epeiph 2.

The many divergences between the two sets of weather data, and in particular the fact that Ptolemy's list includes significantly more dates, some of which tend to cluster around the shared ones, surely point to interventions in the tradition that go beyond accidental dropping out of entries. But fundamentally the separate testimony of Ptolemy supports the hypothesis that what Euctemon compiled was in the first instance a collection of linkages between astronomical and weather events, all set at specific intervals of days. Thus the stellar events have a twofold function, serving both as indices of the stage of the solar year and as weather signs in their own right.

This linkage of the stellar events with contemporaneous weather changes seems on the face of it to have been absent, or nearly so, from the lay employment of such events as date markers discussed in the preceding section, except for parts of the Hippocratic tradition. Hesiod, it is true, speaks of Sirius as "parching the head and knees" (587), but one could plead that Sirius was a special case, supposing the expression must be taken literally. On the other hand, one of the pseudo-Aristotelian *Problemata* (1.3, 859a), reflecting notions circulating among the early Peripatetics of the late 4th or early 3rd century BC, asserts as a matter of fact that winds, waters, and good and bad weather are affected by "the seasons and the risings of stars such as Orion and Arcturus and the Pleiades and Sirius", which is precisely the traditional list of events. Another of the *Problemata* (26.12, quoted at the head of this section) goes so far as to assert a *causal* connection between stellar risings and weather changes, though typically for this genre the text offers an alternative theory that even a weather change coinciding with the rising of Sirius is really caused by the sun.

After Euctemon, the next parapegmatist for whom both "Geminus" and Ptolemy provide a large number of dates is Eudoxus. The first thing that strikes one on comparing the citations of Euctemon and Eudoxus is that Eudoxus used exactly the same set of stars and

Star/Constellation	Specific parts referred to	Magnitude of brightest star
Aries	beginning, entirety	2.00
Taurus	horns. head, tail, entirety	0.85
Gemini	beginning, middle	1.14
Cancer	beginning, entirety	3.52
Leo	beginning, middle, entirety	0.05
Virgo	shoulders, Spica, middle, entirety	0.98
"Claws" (Libra)	beginning	2.61
Scorpio	brow, bright star	0.96
Sagittarius	beginning, entirety	1.85
Capricorn	beginning	2.87
Aquarius	middle	2.91
Pisces	southern fish, node	3.62
Orion	entirety	0.12
Sirius		–1.46
Arcturus		–0.04
Pleiades		2.87 (Alcyone)

Table 2. Stars and constellations appearing in the Callipic entries of the "Geminus" parapegma.

constellations as his predecessor. Again the astronomical phenomena and the weather changes in the "Geminus" parapegma tend to fall on the same dates, though there are more exceptions than was the case for Euctemon (figure 2). And again there is a unique alignment of Ptolemy's weather dates with those of the "Geminus" parapegma such that we get a significant number of coincident dates with substantially the same statements about the weather; in this case one has to equate the summer solstice with Epeiph 1 (June 25), the same date that Ptolemy gave for the solstice in his own time. This alignment also brings a number of Ptolemy's weather dates in line with dates of astronomical phenomena in the "Geminus" parapegma for which there are no coincident weather statements in that document. But Ptolemy also has numerous weather dates that have no counterpart at all in "Geminus," and that cannot all have been tied directly to astronomical phenomena. It is not clear how Eudoxus would have specified such dates, whether as so many days before or after an astronomical phenomenon or in relation to a day count covering the entire solar year. There seems in any case to be a slight, but only a slight, diminution in the independent significance of the stellar dates as something beyond markers of the progress of the natural year.

The "Geminus" parapegma reports a large number of astronomical dates for only one other authority, Callippus. The list of stars and constellations appearing in the Callippic entries is strikingly different from those associated with Euctemon and Eudoxus (see table 2). Except for precisely the four objects found in Hesiod (and this overlap is surely significant), Callippus abandons Euctemon's list, and replaces it with the twelve zodiacal constellations. He manifestly has the constellations in view, not equal divisions of thirty degrees, as is apparent not only from the references to individual features of many of them but also from the irregular intervals separating the dates associated with the beginnings and completions of their risings and settings. It deserves to be remarked that several of the zodiacal constellations are quite as inconspicuous as the dimmest of the asterisms in Euctemon's list.

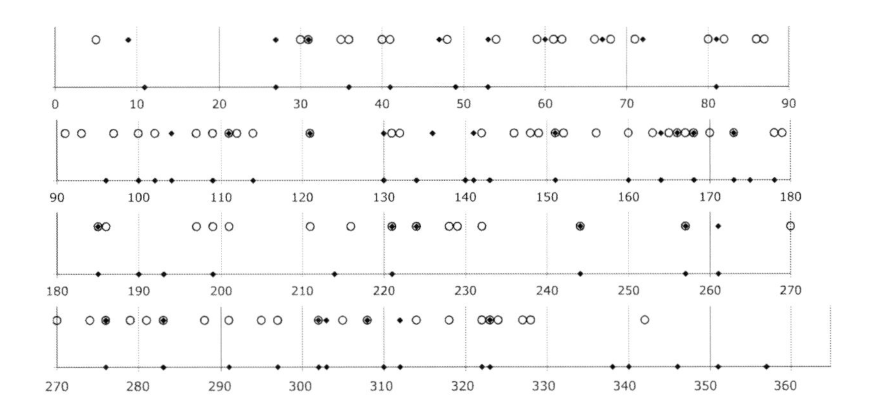

FIGURE 2. Concordance of parapegmatic dates attributed to Eudoxus according to dates counted from summer solstice. Layout as in figure 1.

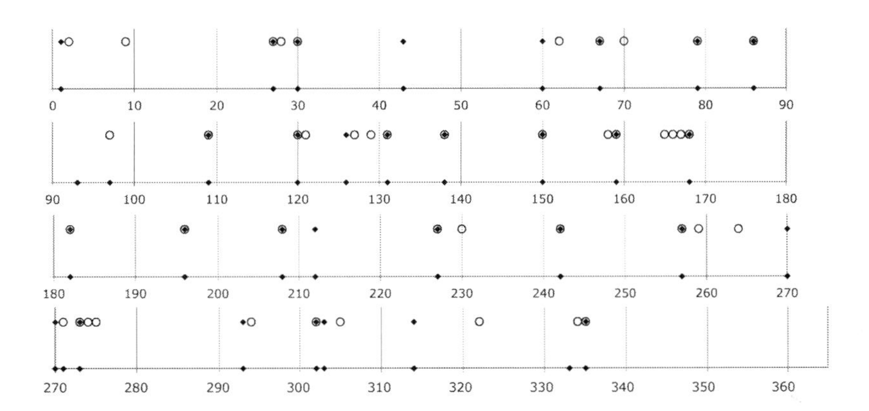

FIGURE 3. Concordance of parapegmatic dates attributed to Callippus according to dates counted from summer solstice. Layout as in figure. 1.

The optimal synchronization of Ptolemy's weather dates for Callippus with those in the "Geminus" parapegma requires an equation of the summer solstice with Epeiph 3 (figure 3). Ptolemy clearly had a very similar, and not much larger, data set for Callippus. The rate of coincidence of weather dates, from either source, with stellar dates in the "Geminus" parapegma is high, indeed higher than in the case of Eudoxus. It is tempting to hypothesize that Callippus was promoting a theory of weather changes according to which the signifying power was almost wholly restricted to the zodiacal belt.

3. Towards a Zodiacal Calendar

> [The parapegmatists] chose a beginning for the year, and took note of which zodiacal sign the sun was in at the year's beginning and at which degree, and they recorded the general changes of the air, winds, rains, and hail that occurred in each day and month, and they set them alongside the sun's positions by sign and degree. After observing this for a number of years, they recorded the changes that happened especially around the same places in the zodiac in the *parapegmata*.... But since they were unable to record a definite day or month or year in which any of these things was brought about because the beginnings of the years are not the same among all peoples and moreover the months do not have the same nomenclature and the days are not counted in the same way among all, they chose to define the changes of the air by means of certain fixed signs.... And just as a beacon is not itself responsible for a conditions of war but rather is a sign of an occasion for war, in the same way the risings of the stars are not themselves responsible for the changes in the air but are set out as signs of such conditions. [Geminus, *Isagoge* 17.7-11.]

> Dionysius named the twelve months, which had thirty days, by transference from the twelve zodiacal signs, and likewise (named) the days from the degrees at which the sun was approximately in mean motion. The first year of his Summer Solstices was in the 463rd year from Nabonassar. [Scholion to *Almagest* 11.3, text 2 in Jones (2003)]

The passage quoted above from Geminus' chapter on parapegmata is a patent fiction concocted to butress his argument that stellar risings and settings do not influence weather, and not the least implausible element of his story is his claim that the early parapegmatists initially correlated observations of weather with solar longitude rather than with stellar phenomena. Geminus' shortcomings as a historian of astronomy ought not to distract us from the fact that his book is a rather good mirror of the range of astronomical practice of his own time, the late Hellenistic period. In the present instance his account reflects what appears to have been the prevailing conventional organizational principle of Hellenistic parapegmata. The "Geminus" parapegma, for example—whose connection with the author Geminus is, as mentioned already, doubtful—does not simply count days from the summer solstice, but divides its year into twelve parts comprising stated numbers of days ranging from as few as 29 to as many as 32, during each of which the sun is stated to traverse one of the zodiacal signs. The near-uniformity of the time intervals, and the regularity of their distribution through the year, make it clear that the text is referring to equal signs of 30° rather than the homonymous constellations. One of a pair of fragmentary parapegma inscriptions from Miletus, dating from roughly 100 BC, was laid out in sections corresponding to the sun's passage through the twelve signs, with the number of days in each section recorded above the series of peg holes and inscribed texts that stand for the individual days and the associated phenomena.[9] Also from about 100 BC we have an artifact that corresponds still more closely to Geminus' story, namely the

Antikythera Mechanism, whose front dial had a revolving pointer displaying the sun's longitude along a dial graduated into zodiacal signs and degrees. Letters inscribed at irregular intervals along this scale served as indices to a list of stellar risings and settings.[10]

The earliest surviving parapegma is a papyrus, *P. Hibeh* 1.27, which can be dated on external and internal grounds to about 300 BC.[11] This text differs from the other Hellenistic parapegmata mentioned above in certain important respects: it organizes its data according to the Egyptian calendar rather than by a true solar year, and in addition to stellar dates and weather changes, it associates with specific dates computed values for the length of daylight in hours as well as numerous religious festivals. Although it does not use the sun's passage through the zodiacal signs to define the chronological framework, it does contain the following entries mentioning zodiacal signs:

> Tybi, in Aries, [day] 20, vernal equinox, the night 12 hours and day 12, and feast of Phitorois. [col. iv ll. 62–64]
>
> Mecheir 6, in Taurus, Hyades set acronychally, the night 11 $\frac{1}{2}$ $\frac{1}{10}$ $\frac{1}{35}$ hours, the day 12 $\frac{1}{3}$ $\frac{1}{45}$. [iv 66–68]
>
> Phamenoth 4, in Gemini, [Capella] rises [in the morning], the night 11 $\frac{1}{45}$ hours, the day 12 $\frac{2}{3}$ $\frac{1}{4}$ $\frac{1}{20}$ $\frac{1}{90}$. [vi 88–90]
>
> Pharmuthi, in Cancer, 3, Sagitta sets acronychally, the night 10 $\frac{1}{3}$ $\frac{1}{30}$ $\frac{1}{90}$ hours, the day 13 $\frac{1}{2}$ $\frac{1}{90}$ [*sic for* $\frac{1}{10}$] $\frac{1}{45}$. [vii 107–109]
>
> Pachon 6, in Leo, Vindemiatrix rises, the night 10 $\frac{1}{4}$ $\frac{1}{30}$ $\frac{1}{180}$ hours, the day 13 $\frac{2}{3}$ $\frac{1}{30}$ $\frac{1}{90}$. [ix 129–132]
>
> Payni [4], in Virgo, Sagitta sets in the morning, the night 10 $\frac{2}{3}$ $\frac{1}{5}$ $\frac{1}{30}$ $\frac{1}{90}$ hours, the day 13 $\frac{1}{15}$ $\frac{1}{45}$. [x 137–140]
>
> Mesore 2, in Scorpio, Pleiades rise acronychally, the night 12 $\frac{1}{5}$ hours, the day 11 $\frac{2}{3}$ $\frac{1}{10}$ $\frac{1}{30}$, feast of Apollo. [xiii 181–186]

These entries have in common that each one is the first entry for its Egyptian month, and each is associated with some stellar phenomenon. The intervals between them, moreover, do not exhibit a regular pattern, although most are within a few days of 30 days (the exception is the first interval from Tybi 20 to Mecheir 6, which is only 16 days).[12] This irregularity is surely to be accounted for by the other facts just mentioned, that is, the stellar dates came first, and the zodiacal references have been attached to them at a secondary stage.[13] Hence I do not think that the interpretation that has been generally accepted since the first publication of *P. Hibeh* 1.27, that these are intended as dates when the sun crosses into the designated zodiacal sign—or alternatively its homonymous constellation—can be correct. Instead, I would suggest that their origin was in a list that simply stated, e.g., that the sun moves into Aries during the month Tybi, and Taurus during the month Mecheir, and so forth. This list was incorporated into the parapegma by the simple, if ignorant, expedient of inserting these statements as part of the first listed event for each month. If this is the correct explanation, then little can be said with certainty about when precisely the passages from one sign to the next were supposed to occur, and in particular whether the sun's passages into Aries, Cancer, Libra, and Capricorn were assumed to coincide with the equinoxes and solstices, as we find in the "Geminus" parapegma.[14] The balance of probability favours interpreting the entries in the Hibeh parapegma as referring to equal zodiacal signs rather than constellations, since this is clearly what the later tradition was doing whereas there seems to be no instance of a Greek text from any period speaking of sequential solar entries into the zodiacal constellations.

The distinctive set of twelve zodiacal constellations appears to have been present already in the *Phaenomena* of Eudoxus, who may in fact have been the person responsible for introducing them as a concept in Greek astronomy. On the other hand, we have no clear evidence that a division of the zodiacal belt into twelve equal signs figured in Eudoxus' work.[15] The current consensus is that the earliest parapegmatists, who were active a generation or more before Eudoxus, would not have organized parapegmatic data by dividing the year according to the sun's passage through the twelve signs.[16] When this practice began is anyone's guess, but the Hibeh parapegma's zodiacal references, garbled though they appear to be, establish the end of the 4th century BC as a probable *terminus ante quem*. I am tempted to suggest that the innovation was due to Callippus, in the light of his preoccupation with zodiacal constellations as significant objects to correlate with weather.[17]

So far as one can tell from the extant parapegmata, the sun's entry into a zodiacal sign was not regarded in its own right as an event to be linked to weather phenomena. What, then, was the purpose of including them in a parapegma? One rather trivial motive is that a year is a cumbersomely large block of days within which to keep track of the current date. Partitioning the solar year into seasons delimited by the solstices and equinoxes and perhaps by the most important stellar dates would be a helpful step, but a division into twelve more or less equal parts, resembling traditional calendar months in length though not tied to lunar phases, would have an obvious attraction, not least because each part would tend to occupy a roughly uniform space in a manuscript or inscription.

Another, more theoretical consideration that might lie behind these zodiacal references is an interest in modelling and predicting stellar visibility dates, such as we can see in the surviving mathematical literature on spherical astronomy. Equal twelfths of the ecliptic, sometimes identified by the names of the zodiacal signs and sometimes not, were regularly employed in Greek "spherics," both as the objects for investigation in their own right in relation to the times in which they rise and set (e.g. in Euclid's *Phaenomena*) and as references for developing theories of conditions of stellar visibility (e.g. in Autolycus' *De Ortibus*). Of particular relevance for parapegmata is Autolycus' modelling assumption, that the condition for a star's visibility is that the point of the ecliptic that rises (or sets) simultaneously with the star should be ahead (or, for setting, behind) the sun by half a zodiacal sign in longitude.[18] We can only guess at the existence of a lost astronomical literature—and how early?—bridging the purely theoretical treatment of visibility in spherics to the bald lists of dates in the parapegmata.[19] For that matter, it is not clear whether the dates assigned to stellar phenomena in the "Geminus" parapegma and others are purely empirical or in part derived from modelling.

Since a parapegma is by its nature an idealization applicable to *any* solar year, the division of the year into twelve sections according to the sun's traversal of the zodiacal signs does not properly speaking constitute a calendar, which would entail reference to specific dates. On the other hand, Ptolemy's *Almagest* preserves eight instances of dates expressed in a true zodiacal calendar that has so far not turned up in any other ancient source. The formulae are uniform in structure in providing a year number "according to Dionysius," a month name derived from a zodiacal sign, and a day number. These were dates of observations of the positions of one or another of the planets relative to fixed stars, and in discussing them Ptolemy provides what he asserts are equivalent dates in his own chronological framework, which uses the Egyptian calendar with year numbers counted from the "first year of Nabonassar" (i.e. the Egyptian year that began in 747 BC). Ptolemy's Egyptian dates can be converted into our Julian chronology. Since, however, the planetary observations took place at night, either after sunset or before sunrise, it is not clear whether Ptolemy understood the Dionysian dates to pertain to the days preceding or following the nights that he specifies (without ambiguity) in his Egyptian reckoning; in other words, we do not know whether he believed that the Dionysian day was considered to begin with sunset, sunrise, or the middle of the night. The following lists the eight Dionysian date formulae with Ptolemy's equivalents translated into the Julian calendar.

Year 13 Aigon [Capricorn] 25, morning	272 B.C. January 17/18
Year 21 Skorpion [Scorpio] 22, morning	265 B.C. November 14/15
Year 21 Skorpion 26, morning	265 B.C. November 18/19
Year 23 Hydron [Aquarius] 21, morning	262 B.C. February 11/12
Year 23 Tauron [Taurus] 4, evening	262 B.C. April 25/26
Year 24 Leonton [Leo] 28, evening	262 B.C. August 23/24
Year 28 Didymon [Gemini] 7, evening	257 B.C. May 28/29
Year 45 Parthenon [Virgo] 10, morning	241 B.C. September 3/4

Ptolemy provides no explanation of the Dionysian calendar and its conventions, and the only supplementary information that we have came to light only in recent years and consists of three brief scholia from the margins of medieval manuscripts of the *Almagest*, one of which is quoted at the head of this section.[20] These texts declare various things: (1) Dionysius lived in Alexandria; (2) the first of his "summer solstices" was in Nabonassar year 463 (i.e. the Egyptian year beginning November 3, 286 BC); (3) the months of his calendar were named for the zodiacal signs, and had thirty days each; and (4) the days approximately corresponded to the degrees of the sun's mean longitude within the relevant sign.

There is no question that Dionysius' year reckoning counted from the summer solstice of 285 BC as the beginning of year 1, and it may be significant for the meaning of this epoch that the Egyptian year that Ptolemy calls Nabonassar year 463 was by convention considered to be the first regnal year of Ptolemy II Philadelphus. Concerning the structure of the calendar, the scholia are unfortunately not consistent, since it is not possible for zodiacal months and days to be coordinated with the signs and degrees of the sun's mean solar longitude if every zodiacal month has exactly thirty days. All attempts to reconstruct the Dionysian calendar have attempted to find a pattern fitting the eight dates in the *Almagest*.[21] The earliest efforts supposed that Dionysius' zodiacal months were unequal in length, resembling the divisions of the year in parapegmata. In the mid 19th century Karl Lepsius and August Böckh showed that no such scheme could be made to be consistent with all the date reported by Ptolemy. Instead, they deduced that the Dionysian year was analogous in structure to the Egyptian year, comprising twelve months of exactly thirty days followed by a supplementary period of five days, but with the year beginning at the summer solstice. Since the Dionysian years exhibit no long term backwards shift relative to the solstices as they would if all years had 365 days, Lepsius and Böckh further hypothesized that a sixth supplementary day was inserted at four-year intervals. Thus the Dionysian calendar would seem to have anticipated the reform of the Egyptian calendar that Ptolemy Euergetes abortively proposed in the

"Canobic Decree" (*OGIS* 56) in 239/238 BC and that was finally adopted at the beginning of Roman rule in Egypt.

The reconstructed Dionysian calendar of Lepsius and Böckh fits Ptolemy's equations of Dionysian and Egyptian dates almost perfectly, though there is a single discrepancy of one day. However, Lepsius and Böckh were unaware that not all the planetary observations that Ptolemy reports describe what should have been seen on the nights that he associates with them: one of them seems to be dated one or two days too early, and two others seem to be one or two days too late for the situation he describes the planet as being in.[22] Now internal consistency guarantees that the wording of the reports, and the Egyptian dates that Ptolemy assigns to them, have been transmitted accurately as he wrote them. The inconsistencies must be due to one of four causes: (1) the original observers may have inaccurately described what they saw (though this is hard to believe in the present instances), (2) Ptolemy's version of the reports may not correctly reproduce the originals, (3) the Dionysian day numbers in our manuscript tradition of the *Almagest* may be corrupt, or (4) Ptolemy's conversions of the dates from the Dionysian to the Egyptian calendar may not be reliable. Of course more than one of these sources of error may apply; but the high frequency of discrepancies suggests a common explanation, and the most plausible one is that Ptolemy had an incorrect or only approximate rule for converting between the calendars. And interestingly, if we replace the problem dates with dates that better fit the observation reports, the equations are no longer consistent with an Egyptian-style division of the year into thirty-day months, but suggest a parapegma-style division into zodiacal months that are shortest in the autumn and longest in the spring.[23] Enough uncertainties remain to prevent us from offering a complete reconstruction of the calendar; in particular it would seem that the summer solstices marking the beginnings of the years cannot have followed a rigid four-year cycle of 365-day and 366-day intervals. Nevertheless this interpretation of the Dionysian dates makes good sense in the context of contemporary astronomical practices. By contrast it is difficult to understand why one would have adopted a calendrical structure according to which the months progressively run ahead of the sun's passage through the homonymous zodiacal signs, accumulating a lead of five days by the end of the year that nominally belongs to none of the signs.

If, as seems likely, Dionysius was himself active as an astronomer during the inaugural year of his calendar, then the calendar had a career of at least forty-five years, which is probably too long for a single astronomer's career; on the other hand there is no reason to suppose that anyone outside a small localized team of observers used it. The purpose of their observations is anyone's guess—were they investigating the patterns of planetary motion in their own right, or tracking supposed correlations between these motions and mundane phenomena? In any event, the Dionysian calendar failed to catch on among later Greek astronomers, perhaps because it fell between two stools: the Callippic calendar offered a more natural mapping on to the calendars in local use throughout Greece and the Near East, while the Egyptian calendar was simpler and more regular, and thus suited to long term astronomical calculations.[24] In due course Roman power accomplished what the astronomer Dionysius could not, imposing upon the eastern Mediterranean region a family of calendars modelled on Caesar's Roman calendar with its solar years, albeit divided into months of no astronomical significance whatsoever.

Notes

1. Samuel (1972); Trümpy (1997).
2. The scholarship on the Athenian calendar is vast, and it is difficult for the neophyte to strain the secure facts out of a brew of a priori conjectures, tralaticious assumptions, and polemics. A patient reader can trace a substantial bibliography by working backwards from the references in Pritchett (2001). The thesis presented in Pritchett & Neugebauer (1947) (and maintained in Pritchett's many subsequent publications on the subject) that the Athenian calendar was subject to irregular and capricious intercalations of months and insertions and compensating deletions of days has on the whole held up well against a steady barrage of rival models. Müller (1991) and (1994) infers from numismatic evidence that the Athenian calendar was intercalated according to a 19-year cycle from the 2nd century BC; J. D. Morgan's often-cited researches on Athenian chronology, which also entail an effective 19-year cycle of intercalation already in the early 4th century BC, were endorsed by Habicht (1997), pp. v–vi but unfortunately remain unpublished (cf. Morgan (1996) and (1998)). Presuming his results are correct, it remains unclear whether the Athenian year's length was ever consciously regulated by a cycle (which ought then to have specified which months were repeated as well as the dates of summer solstices) or whether the cyclic pattern was an automatic consequence of starting the new year consistently with the first new moon following an independently determined summer solstice.
3. Jones (2000) and literature cited therein. The hypothesis that the Athenian month names (accompanied by Athenian archon-names for the years) in the reports of three 4th century BC Babylonian lunar eclipse observations cited in Ptolemy, Almagest 4.11 pertain to a putative calendar devised by Meton is plausible enough but cannot be confirmed.
4. The Homeric poems refer by name to a few constellations but contain little else that counts as astronomy.
5. The most concrete evidence for the date of writing of the Epidemics is the inclusion of case histories from Olynthos in Books 5 and 7, which situates these books before the destruction of the city in 348 BC. The other books are generally presumed to be several decades older than this pair.
6. Lehoux (2007).
7. On Euctemon's parapegma see Hannah (2002).
8. Hannah (2005), pp. 62–70 suggests that the density of stellar dates in Euctemon's list may have been motivated by the varying seasonal activities, agricultural and other, of a Greek community. I am not quite convinced that the pattern of gaps and dense patches in figure 1 bears this out.
9. Lehoux (2005).
10. Price (1974), pp. 18 and 46–49.
11. Grenfell and Hunt (1906), pp. 138–157.
12. The original editors of the papyrus tried to remove this anomaly by conjecturing an omitted day number (5) following "Tybi," effectively turning one entry into two separate ones; this editorially inserted day number has been treated as if it was actually in the papyrus in some of the more recent scholarship. The day numbers in the papyrus are all guaranteed to be correct by the associated figures for the length of night and day, which conform to an arithmetical scheme.
13. Spalinger (1991) proposes a different, rather complicated hypothesis by which a scheme comprising only 30-day and 31-day zodiacal intervals could have been distorted in the transition from a Greek parapegma to a parapegma structured on the Egyptian calendars (both civil and lunar). The stellar phenomena play no role in his reconstruction.
14. If the dates in the papyrus are to be taken seriously as the actual passage dates, then no consistent alignment of the solstitial and equinoctial points in the signs is possible, but all would be at or past the middle of their respective signs.
15. Bowen and Goldstein (1991), pp. 241–245. Aristotle, Metaphysica 1073b, appears to be the first extant author to speak of the ecliptic circle as the circle "through the middle of the signs [dia mesôn tôn zôidiôn]," in the context of his summary of the homocentric sphere models of Eudoxus, but one cannot be sure whether zôidion here meant equal signs or constellations. In 3rd century BC authors such as Autolycus and Aristarchus zôidion definitely means a twelfth of the ecliptic or by extension a twelfth of a great circle in general.

16. Hannah (2002), pp. 120–123.
17. It would be gratifying to adduce as evidence of Callippus' part in incorporating zodiacal elements into the parapegma tradition the scheme of intervals for the sun's passage through the signs published by Rehm as Callippus', for which see Rehm (1949), 1346–1348. Unfortunately I share Neugebauer's scepticism of this as of so many of Rehm's reconstructions of parapegmatic schemes on the basis of considerably later sources (Neugebauer (1975), pp. 628–629).
18. Neugebauer (1975), pp. 761–763.
19. The date commonly assumed for Autolycus, c. 300 BC rests on Diogenes Laertius' report (4.29) that the Academic philosopher Arcesilaus studied with him in his youth; for reservations about the trustworthiness of this story, see Bowen and Goldstein (1991), p. 246 n. 29.
20. The scholia were first published in Jones (2003) as Texts 1–3.
21. For details of the historiography of the Dionysian calendar, with references, see Jones (2006), pp. 284-287.
22. Jones (2006), pp. 258–270.
23. Jones (2006), pp. 287–289.
24. Ross (2006) interprets certain problematic uses of zodiacal signs in Demotic horoscopic ostraca from Roman period Medinet Madi as survivals of the Dionysian calendar. Supposing that the dating formulae in these ostraca have been correctly understood, an independent reinvention of zodiacal dating seems more plausible than a line of transmission undocumented over more than three centuries.

References

Bowen, A. C., and Goldstein, B. R., 1991, "Hipparchus' Treatment of Early Greek Astronomy: The Case of Eudoxus and the Length of Daytime", *Proceedings of the American Philosophical Society* 135, 233–254.

Grenfell, B. P., and Hunt, A. S., 1906, *The Hibeh Papyri* (London: Egypt Exploration Fund).

Habicht, C., 1997, *Athens from Alexander to Antony* (Cambridge, Massachusetts: Harvard University Press).

Hannah, R., 2002, "Euctemon's Parapegma", in C. J. Tuplin and T. E. Rihll (eds.), *Science and Mathematics in Ancient Greek Culture* (Oxford: Oxford University Press), 112–132.

——, 2005, *Greek and Roman Calendars. Constructions of Time in the Classical World* (London: Duckworth).

Jones, A., 2000, "Calendrica I: New Callippic Dates", *Zeitschrift für Papyrologie und Epigraphik* 129, 141–158.

——, 2003, "A Posy of Almagest Scholia", *Centaurus* 45, 69–78.

——, 2006, "Ptolemy's Ancient Planetary Observations", *Annals of Science* 63, 255–290.

Lehoux, D., 2005, "The Parapegma Fragments from Miletus", *Zeitschrift für Papyrologie und Epigraphik* 152, 125-140.

——, 2007, *Astronomy, Weather, and Calendars in the Ancient World* (Cambridge: Cambridge University Press).

Morgan, J. D., 1996, "The Calendar and the Chronology of Athens" (abstract), *American Journal of Archaeology* 100, 395.

——, 1998, "Polyeuktos, the Soteria, and the Chronology of Athens and Delphi in the Mid-Third Century B.C." (abstract), *American Journal of Archaeology* 102, 389.

Müller, J. W., 1991, "Intercalary months in the Athenian dark-age period", *Schweizer Münzblätter* 41, 85.

——, 1994, "Synchronization of the Late Athenian With the Julian Calendar", *Zeitschrift für Papyrologie und Epigraphik* 103, 128-138.

Neugebauer, O., 1975, *A History of Ancient Mathematical Astronomy* (New York, Heidelberg, Berlin: Springer-Verlag).

Price, D. de S., "Gears from the Greeks. The Antikythera Mechanism: A Calendar Computer from ca. 80 B.C.", *Transactions of the American Philosophical Society* N.S. 64.7.

Pritchett, W. K., 2001, *Athenian Calendars and Ekklesias*, ΑΡΧΑΙΑ ΕΛΛΑΣ 8 (Amsterdam: J. C. Gieben).

——, and Neugebauer, O., 1947, *The Calendars of Athens* (Cambridge, Massachusetts: Harvard University Press).

Rehm, A., 1949, "Parapegma", in *Realencyclopädie der classischen Altertumswissenschaft* ("*Pauly-Wissowa*") 18.4, 1295–1366.

Ross, M., 2006, *Horoscopic Ostraca of Medînet Mâdi*, Ph.D. Dissertation, Brown University.

Samuel, A. E., 1972, *Greek and Roman Chronology. Calendars and Years in Classical Antiquity*, Handbuch der Altertumswissenschaft I, 7 (München: Verlag C. H. Beck).

Spalinger, A., 1991, "Remarks on an Egyptian Feast Calendar of Foreign Origin", *Studien zur altägyptischen Kultur* 18, 349–373.

Trümpy, C., 1997, *Untersüchungen zu den altgriechischen Monatsnamen und Monatsfolgen*, Bibliothek der klassischen Altertumswissenschaften N.F. 2. Reihe Bd. 98 (Heidelberg: Universitätsverlag C. Winter).